DATE DUE

S

Architecture
IN AN AGE OF SCEPTICISM

Architecture
IN AN AGE OF SCEPTICISM

A Practitioners' Anthology Compiled by
Denys LASDUN

Oxford University Press
New York 1984

To S.V.L.

First published in Great Britain by William Heinemann Ltd.,
London, 1984

First published in the United States by Oxford University Press,
Inc., New York 1984

Library of Congress Cataloging in Publication Data
Main entry under title:

Architecture in an age of scepticism.

  (Cinema classics)
  Includes index.
  Contents: Sketches of a new architecture/Christopher
Alexander – Nursed by experience/Edward Cullinan –
The University Centre, Urbino/Giancarlo de Carlo –
[etc.]
  1. Architecture, Modern – 20th century – Addresses,
essays, lectures. I. Lasdun, Denys. II. Alexander,
Christopher.
NA680.A6914   1984   724.9'1   83-43273

ISBN 0-19-520445-X

Designed by Dennis Crompton
Printing (last digit): 9 8 7 6 5 4 3 2 1
Printed in Great Britain

CONTENTS

Sunday evening in the Piazza San Marco; an image evoking a sense of time, place and people – people engaged in creating space and form, a microcosm of the city. This picture has been a continuing inspiration to me for over a quarter of a century. Regrettably I can neither remember, nor discover, its provenance or authorship. I can only offer to the unknown photographer my thanks and admiration. DL

PREFACE

When the common values of society are generally in doubt, as they are today, and a feeling of scepticism increasingly prevails, architects have an obligation to speak for themselves and to explain what architecture means for those who have actually designed and built the buildings, who have kept faith with and continued the essential living tradition and are daily concerned with the practice of architecture. Doubt and cynicism and uncertainty need to be met by a strong affirmation of those positive ideas in which architects can and do believe and by which they are able to create. This book is meant to say what architects alone can say; it aims to present the understanding of the architect rather than the analysis of the critic or the historian; its thought is *with* rather than *about* architecture. It aims to offer a unique impression of the way architects' minds work and so to be useful for architects, students and laymen alike. Whatever direction architecture may take and whatever the immediate future may bring, the testimony of those who have borne the brunt of the post-war period will remain crucial. They have much to hand on to their successors; we all have to evolve and develop but, in doing so, we must preserve intact those creative certainties which make architecture possible.

It may be said that architects ought to stick to building and leave writing about architecture to others, but there is a long tradition of key volumes on architecture written by architects stretching if not from Vitruvius, then at least from Alberti, to Le Corbusier and beyond. This book – a practitioner's anthology – aspires to a modest place in the series. Whatever the contributors may lack in the way of literary finesse, they make up for by their direct experience, their unique insights and their concern for the essence of architecture. They do not form a particular group and are not motivated by any single set of ideas. The selection is neither exclusive nor exhaustive. The contributors are, I think, all individuals who are not uncongenial to each other – Europeans who work in the tradition of the modern movement. They have all produced buildings full of eloquence and human depth. Their works cover a wide range of architectural endeavour from a simple table to the complexities of a city and are intended for everyone, from the highest powers of the state and the most rarefied organs of learning to the great urban and agricultural majority of the human race.

Each contribution gains clarity and power from the company of its fellows; the effect of the essays is cumulative and shows an impressive range of mature buildings that modern architecture – in the broadest sense – has managed to create in the last half of the twentieth century.

All the contributors have a similar seriousness in their approach to architecture, a similar desire to express the fundamentals of their art and to advance it in its relation to people and to society. They do come close to each other when they express what architecture is about for them. *'Architecture is the art of communities.'* *'The business of architecture is to establish emotional relations by means of raw materials.'* *'I began to search for ways of building which would allow some kind of genuine spirit to show itself.'* *'The most ideal conditions for human beings that can be created out of the programme and the means given to the architect.'* *'To explore such intangibles as proportion and the appropriateness of celebrating a handsome structure by expressing it honestly.'* *'We always find ourselves trying to create an entity which is both part of a larger one and a grouping of smaller ones.'* *'Architecture can do no more, nor should it ever do less, than accommodate people well – assist their homecoming.'* *'The rooms flow into one another freely like the compartments of a cave and, as in a cave, the skewed passage which joins the compartments effectively maintains privacy.'* *'A building's completion takes many years because it has to be nurtured by the transformations produced by use and the processes of integration set up between the artefact and nature, between building and environment.'* *'There has been a return to the more ancient desire to see buildings wherein a primary objective is for them to appear appropriate in their context.'* *'What remains common to this kind of architecture is that it operates through the central medium of formal ideas built up around problems that have to be solved. It is not concerned with fashion.'*

There seems to be a common voice here; a concern with a sense of time, place and people, and with the way space must be organized if architecture is to play its part in creating and maintaining the well-being of human society. We need to make it easier to understand how architecture really is created today without trying – as is so often done – to marshal it into categories and to make it a branch of history rather than of art.

Although the contributors deal with their work in many different ways, they all try to define the ideas and feelings behind their architecture and do not confine themselves purely to a functional approach although this is clearly the basis of all architecture. They have their own interpretations but within a shared tradition and often show a more relaxed attitude to the relationship between form and function than their forbears. While photographs can give some indication of the way the buildings look, they are no substitute for actually experiencing them. It is not enough just to look at a building; you must move round and through it as its own organization demands. I think all these buildings engage heart and mind and senses and speak with subtlety and power on many levels of meaning. Together they show much of the true nature of the significant architecture of our time, and whilst there is much more to architecture than seductive external finishes, it is nevertheless a sign of our sceptical age that comparatively little is spent on the external fabric of buildings and even less on their maintenance.

I am very pleased that, without exception, all of those whom I invited have been able to find time in their busy lives to write such illuminating explanations of their work.

Finally, I want to thank my colleague Graham Lane for his great help in the planning and production of this book, J.H.V. Davies, whose advice and assistance on all aspects of the project have been invaluable, Dennis Crompton, for his skill and care in the design and my son James for many perceptive suggestions.

Denys Lasdun
1983

Christopher ALEXANDER

SKETCHES OF A NEW ARCHITECTURE

I finished my architectural training (so-called) in 1958. In that year, I came to the conclusion that what I had learned at school, and the prevailing architectural wisdom, meant almost nothing . . . and that, fundamentally, and quite simply, we in our time just did not know how to build buildings. The art of building, as a serious thing, seemed to me to be entirely lost . . . and the original reasons I had for wanting to be an architect – wanting to make buildings which took their place in the great line of buildings, built over the centuries, . . . was impossible to achieve, without a drastic reorganization – indeed, without a definition from scratch . . . of what the art of building really meant.

For this reason, instead of joining an architects' firm, I simply began to think and write – first in my PhD thesis at Harvard . . . and then for years afterwards, in my teaching capacity at the University of California. With one exception – a small school I built in India in the early 'sixties – I intentionally built nothing for almost fifteen years – because I felt that it could not be done well, and I did not know how to do it.

Only, finally, in the early 'seventies, after years of thinking and preparation, did I feel ready to try some experiments.

But, when I did try to put the results of my thoughts into practice, in a series of buildings built during the early 'seventies (the Modesto clinic, and the houses in Lima, among others), these buildings still had the same dead characteristics, typical of the age in which we have all been building.

At that stage I began a second series of experiments, far more radical, and far more difficult to implement. In these experiments I first assumed that I myself would be the builder – the contractor – and began to make the changes in my life, and the necessary preparations for this kind of task. And then, secondly, I began to search for ways of building which would allow some kind of genuine spirit to show itself. Thus, in these experiments, I had the fundamental attitude that I was not designing buildings – but *making* them. And all my work turned towards this act of making, and its art.

Now, this is very easy to write down; but immensely difficult to do. It has taken almost ten years of work, experiments, and projects, to reach the stage where I can really undertake this kind of work, on a reasonably large scale. And the various cases where I have been able to carry something out fully, in the way that I believe is right, are necessarily very tiny.

For this reason, my forthcoming book is called *Sketches of a new architecture*. What I have accomplished so far, are just tiny fragments of what I hope one day to achieve, and of what I hope must be achieved, in the coming decades, by all of us concerned with building.

Of course they are three-dimensional sketches – fragments of actual physical reality – not drawings – but they are sketches nevertheless . . . partly because they are so small . . . and also, because they are in many cases, unfinished, hinted at . . . not yet fully matured, not perfected.

It would be easy to dismiss these small sketches as insignificant. They are tiny, indeed, compared with the huge task of building millions of cubic feet of buildings, every year . . . as we are called upon to do . . . and as the order of the world requires.

But, in order to reach the depth of changes which are reflected in these sketches, it would have been impossible – quite impossible – to make them any larger. In many cases, the physical technique, needed to make one of these sketches – was hardly developed – and could only be carried out in miniature, to see if it worked *at all*. So, in part, it would have been physically impossible to carry out these experiments on a larger scale.

But it would also have been socially, and emotionally impossible. In order to make buildings, in this way, it is necessary to change, completely, our conception of the architect. The social arrangements, contracts, assumptions, obligations, definitions of responsibility . . . which define the architect, the planner, and the contractor, in modern society, are so well established, that it requires enormous force, even to make these changes on a small scale. Some of the anecdotes that accompany the individual cases, will make it clear, how quickly, the means required to implement a given work, were at odds with the methods of implementation available to carry them out.

These fragments, these buildings, and other buildings of this kind, cannot be built by making drawings . . .

This article is a summary of a forthcoming book, which will also be called *Sketches of a new architecture*.

and then handing over to the bank and to the general contractor. They require a different order of society to execute them. And this order of society will be hard won.

Nevertheless, all my experiments, now, are aimed at showing how these fragments can be reproduced, in essence, on a much larger scale. We are now building a small university in Tokyo; a ten-storey apartment building in Sapporo; a village in Israel; and we are responsible for the construction of several thousand houses in a new town in Venezuela. In each of these cases, we have a client, who understands, to some level, what kinds of changes are necessary, in the fabric of procedures, to allow these sketches which are presented here to come to life on a larger scale. If we manage that, then it might truly be said that a new age of architecture has arrived.

BLACK AND WHITE TERRAZZO FLOORS, 1980

In 1980 I spent a few hours in Florence, and was astonished by the intense beauty of the very simple black and white marble floors in the Baptistry, in San Miniato, and in other churches. These floors, laid by *intaglio* work, small chips of black and white marble, achieved a depth and simplicity quite out of proportion with the difficulty involved.

They stayed in my mind, and when I came back to Berkeley, I decided that it must be very simple to find some modern way of producing a similar kind of ornament, and we began a series of experiments in our yard.

Since we were doing all kinds of experimental concrete work, I decided that some form of coloured concrete would be a suitable way of doing it . . . and we soon found that ordinary terrazzo – a mixture of coloured cement, and marble dust and fine marble chips would give us the material we needed. This material is laid wet, then ground to a first finish about twenty-four to forty-eight hours after it is laid, while still green, and then finally polished with a high-speed sander a few days later.

The question was, how to get the pattern, without creating enormous labour cost. I did not want to produce something which was inherently so expensive that it could only be a luxury. My goal, from the outset, was to produce a kind of floor which would be reasonable in price, so that we could build them easily in different buildings, and yet provide a process which can allow the kind of personal feeling which is inherent in this kind of work.

We began by making a simple brass mould, that would allow us to fill first the black material by using half the cells of the mould, and then, after this had set, twenty-four hours later, to fill the remaining space with white material . . . then grind the two together . . . and polish the result.

The first result was good, but tearing at the edges disturbed the pattern. My fears that the result would be too mechanical were allayed. We then built a larger mould of the same type, with a very complex negative 'spider', that kept the material pressed down, while the mould was lifted up. This mould allowed us to make a larger area of the same design, which you can see on the lower step in the photograph.

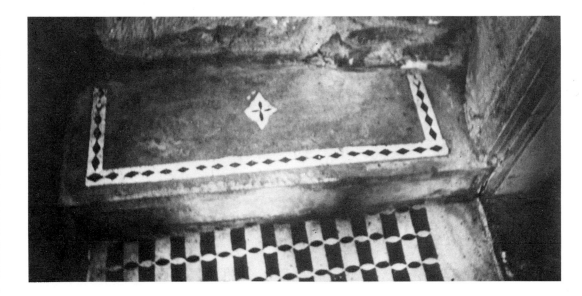

However, the needs of the material are much more fluid. It is necessary, in order to follow the centering process, to have extreme liberty in creating new designs, that are just right, for the place they go into. . . . To see what we wanted on the step above, we first played with black and white paper, to find the right design . . . and then used an entirely different technique to make the physical thing. We made the white part of the design in thin styrofoam – leaving gaps for the black. Then we filled the black terrazzo around the styrofoam. Then, after twenty-four hours, we burnt out the styrofoam with a torch; and then backfilled with the white terrazzo; twenty-four hours later ground it off, later polished and sealed the surface with a shiny sealer.

The new technique is far more beautiful than the old. Even though the first example has a roughness which is not really mechanical, it is still impersonal and mechanical when we compare it with the second, the smaller design on the step, which does somehow reach a spiritual quality . . . because it is so personal. The styrofoam allows the exact shape which the personal vision of the place has in it, to be produced, to the nearest millimetre, exactly as it is felt to be right . . . and this brings the thing to life.

We shall try some larger floors now, in the styrofoam technique.

RESTAURANT AND CAFES ON THE HUDSON RIVER, 1979

In 1979 Bob Schwartz of Tarrytown, New York, asked us to design a huge restaurant on the Hudson river, a kind of place with hundreds of different rooms, cafés, a place to dance, places to entertain privately, almost a kind of Tivoli in a building.

The project went as far as the building of this model: and then came to a halt. Perhaps it will still be built; one day. In any case, even in model form, it is one of the buildings I like best of all those which are represented among these fragments and sketches.

The long awning along the river, which is a beautiful rich red, is itself almost 300 feet long.

TARRYTOWN CONFERENCE CENTER, 1979

Bob Schwartz also owns a piece of land on the Hudson river, where two of the old New York Estates have been joined to form a conference centre. So far, the two houses are however still quite separate, and the area between them, although potentially beautiful, does not join them in any obvious harmonious way. We spent a week with Bob, to help him lay out a kind of dream world, a nucleus of a conference centre, which would unite the two old houses, and form a place between them, that would encourage reflection, study...

The two existing houses do not appear on the drawings. One is above, left, and the other is below, left. The land slopes between them ... and the whole thing is on top of a hill which overlooks the Hudson river valley, with beautiful views in all directions.

The main building we designed stands at the junction of the two old properties. In front of it, is a quiet, water-filled lagoon, surrounded by walkways, with a terrace above it, arcades along the east, and the main building above the terrace, built into the tall trees which are standing there.

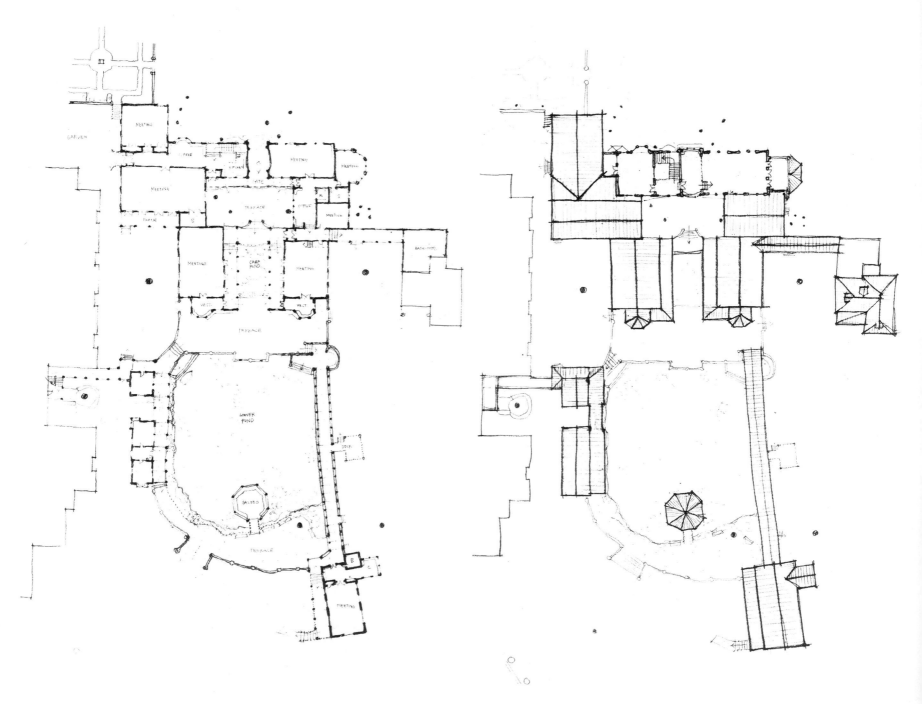

ETNA STREET COTTAGE, 1973

This building was built in ten weeks, without a building permit, at the back of our office on Etna Street in Berkeley. I have described the process of its design fully, in the *Timeless Way of Building*, Chapter 22. It was the first time that I tried, explicitly, to build something according to the construction patterns in *Pattern Language* and also the first time that I had myself taken on the construction of a building as a contractor.

As you can see, the building is very naive and innocent. It arose directly from the plan, following only the rule that we placed columns at the corners, and made the ceilings as vaults, over woven baskets. There was never, at any moment, in my construction of this building, a conscious attempt to define what the building would look like.

SKETCH OF A PAINTING ON WOOD, 1980

In the spring of 1980, I gave a seminar on colour, where I studied, with a small group of students, in which we tried to find the rules which govern the formation of 'inner light' in colour – an extension of the ideas on geometry which I had been formulating since 1976, and which will appear in a later volume of mine.

In order to understand these rules, and to put them to the test, we each of us made small paintings, week after week, in order to show both what result the rules we had formulated tended to produce, when, and in order to find examples of inner light, intuitively, as a way of trying to define other, as yet undefined rules, which might contribute to the effect. This sketch, shown here only because there are no colour reproductions, is a first sketch of one of these paintings.

I include it here, because it sheds light on a general feature of making and building, which lies consistently

through these pages. Even this painting has some quality in it, which seems 'not quite of our time'. It is something that goes beyond the sterile, fashionable, and conceptual art of the twentieth century, and reaches forward, or backward, into something much closer to a more childish human heart.

It seems almost ridiculous to say this, in such a 'sweet' fashion . . . and yet there is a quality of some kind, even here which expresses very clearly what I am trying to show through this piece . . . and which is very, very hard to put into words.

The phrase which comes closest, is that, in some way I am not at all, trying to be clever. These paintings are perhaps the least clever, of any paintings that one can find in the twentieth century – even less clever than the so-called primitive paintings. They have the heart of a simpleton in them; at least the kind of simpleton that sees through to the truth now and then.

HOUSES IN MEXICALI, 1976

These buildings, part of our project in Mexicali in the north of Mexico, are completely described, with colour pictures, in the book *The Production of Houses*, Oxford University Press, 1983.

This was the first large project where I myself was not only architect, but also contractor. We built the houses, with the help of the families, who also laid out houses for themselves, using the pattern language. Our production process covered everything. We even produced our own special earth cement concrete blocks, in a factory, that was right next to the building site. The roofs of the buildings are all thin concrete shells, hand applied over lightweight baskets woven from wood strips, and then covered with burlap and chicken wire.

In 1976, the houses, of about 70 square metres in area, cost about $3,500 each.

COURTYARDS FOR THE CANARY ISLANDS, 1973

This is a sketch, made during a project in which Ingrid and Halim and I undertook the preliminary design of a tourist development in the Canary Islands. The project itself would have been beautiful: people coming to the place, themselves helping to build it, as part of their vacation, great gardens full of flowers, the black rocks of the islands, the yellow sand dunes, a tourist resort where one came to experience a new vision of life, food under the desert night sky, sleeping on the roofs of houses, under the night sky, eating from vendors in the narrow streets, working with the other visitors to tend the gardens, and to build the buildings...

During this project, I made this drawing to show the members of our team, how it might be possible for a pattern of courtyards to grow gradually, and what sort of geometric character they would have, if they grew naturally. I made the drawing, mainly, to contrast this naturalness with the artificial, mechanical character, that courtyards made in our time, typically have.

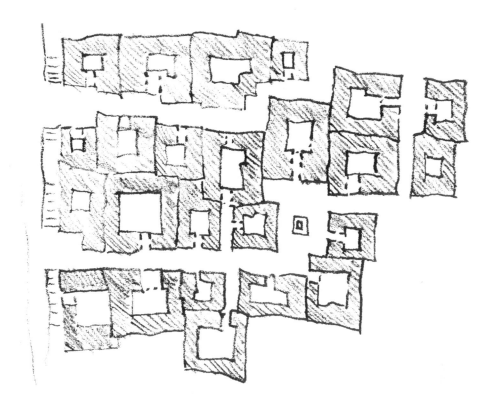

This naturalness comes from five things:
First, the courtyards grow: the process of drawing makes one courtyard at a time, just as they might be made in the actual building process.
Second, as a result, each courtyard is unique, because it sits in a slightly different relation to its neighbours, it plays a different role in the whole, and this shapes it and makes it unique.
Third, all the courtyards follow the same basic rules (the same range of sizes, for the interior, the same range of thicknesses or depths for the buildings, the same passage which goes into each one ...)
Fourth, each one delights in some peculiarity, and makes something of it, so that whatever difference it has, is turned to advantage, made into something.

Fifth, each one is made with great care, care based on its own peculiarities, so that each part of it, every corner, is made whole, while the array is growing, and there are no left-over corners, no dead bits, every part is made whole, and made to come to life.

Even though this is only a pencil sketch, you can see all this, in the drawing, and can imagine how this would happen in the real building process, provided that the real process had the same five rules to follow, as the drawing does.

BRICK FLOOR IN AN ARCADE, 1976

Just bricks. But compare them with the bricks of a mid-twentieth century building made within the architectural establishment. There they would be perfectly regular. Here they are placed in the sand, and you can feel the hands that put them there. There they would be placed with perfection, in their collinearity, each one lining up perfectly with the ones next to it. Here they are placed with perfection in the eye, they are placed so as to make the edge nice, and the bricks at the edge are also cut to make it nice, so that the whole feels right.

There are no bizarre and funky patterns in the bricks. There is no attempt to make this more than it is, no hint of a hunger so great that it has to feed itself by making designs everywhere, with brick designs, brick squares, alternating brick zigzags, brick medallions, brick edges, bricks on their sides, bricks on their ends.

Here there are just hundreds of bricks, all laid in the same way exactly, like the scales of fish, but happy in their regularity, because it comes from ordinary affection.

THE LINZ CAFE, 1980

This small building, built in 1980 on the banks of the Danube, in Linz, Austria is fully described in a book of the same name, *The Linz Café*, Oxford University Press, 1981.

In this building, for the first time, I began to find a way of reaching the kind of informal simplicity, which is indicated in the pattern language, but coupled with a simple coherence of structure, a very formal, but simple layout and design. In this sense, the building is not funky, and moves much further in the direction of pure form than was ever suggested by the work which went into the pattern language.

Perhaps the most important thing about the building (unfortunately impossible to reproduce here) are the glowing colours, and paintings on the walls, apricot yellow, very pale green, deeper green, a more orange version of the apricot yellow, small touches of a deep crimson red . . . all mixed by hand, and painted by hand, in the last two weeks before the building opened.

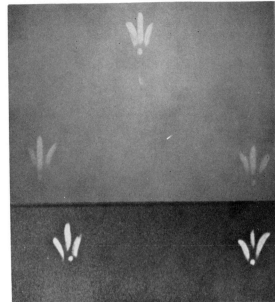

FIRST SKETCH OF THE MARTINEZ HOUSE, 1977

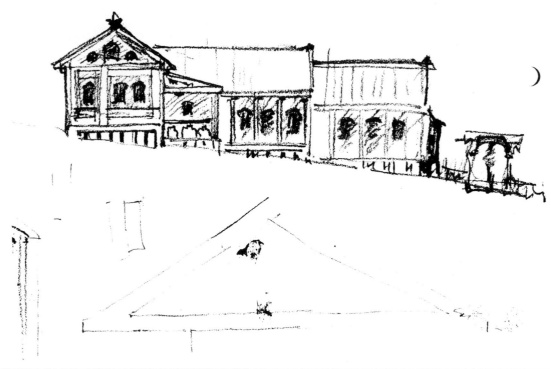

I made this drawing the day we came home from the site in Martinez, immediately after we had laid the building out.

   The plan was still vivid in my mind. I tried to sketch, in a flash, what the building seemed to be like in its outward character.

   Below you can see part of the finished building, and the way it looked when we began to build it . . . but here, there is a substance, more fleeting, harder to catch, but something very simple and ordinary. Especially, the end wall on the left has something. It is so ordinary, it almost begins to be strange . . .

SALA HOUSE, KITCHEN FLOOR, 1982

The Sala House, another house now under construction, has a large farmhouse kitchen, with high ceiling, fireplace alcove, and a forest of tall columns, with wooden corbels holding the beams. We are just about to put in the floor and are going to try our black and white terrazzo technique, for the first time on a large scale. Perhaps it will be dark blue and white in this particular case. These are early sketches of the way the pattern might be made.

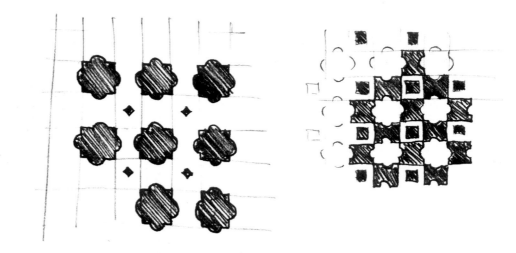

FOUR-STOREY APARTMENTS DESIGNED BY TWENTY-SEVEN FAMILIES, 1974

One of the reasons why the *Timeless Way of Building*, requires that people design their own buildings for themselves, is simply that they know more about their lives, have the capacity to make things just right, need the touch of a personal world around them...

But there is another reason too, which I have never mentioned explicitly in any of my other books. At a time in the twentieth century, when architects are obsessed with images, with their own creative power, with the uniqueness of their designs, families are never concerned with these things ... they are concerned, simply, to make sure that things work.

As a result, when people use the pattern language, to lay out buildings for themselves, there is, inevitably, an informality, an egolessness, similar to that achieved in traditional cultures – just because no one is trying to 'prove' anything, just going about their business to try to get things right.

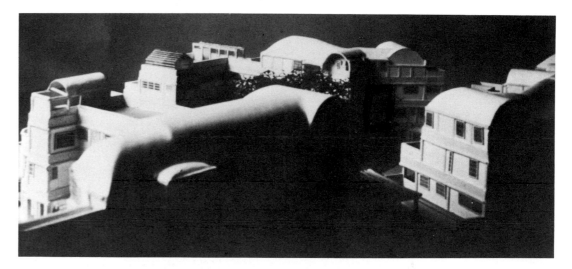

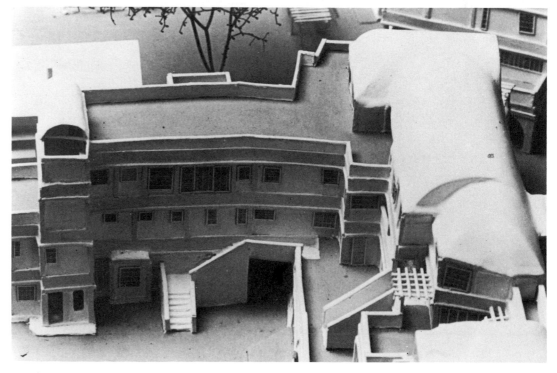

BEDSIDE TABLE, 1971

This little table was made for love. I made it for Ingrid, one Saturday morning between nine in the morning and lunchtime — and gave it her that afternoon as a surprise.

I made a drawing of it on a piece of old wood, went to buy the planks — ordinary pine — came back home and started to cut them at once. It is only glued and nailed, so I put it together almost as fast as I am writing this.

It was made as a gift, a private thing — for her to have beside her bed — and she still has it there to this day.

One of the things I like about this table is its extreme simplicity of construction. It is made entirely out of three-quarter inch planks, nailed and glued to form a box-like structure which becomes rigid because of the three dimensions.

In this case, more than almost anything else in these sketches, I made the table without any idea in my mind that anyone would ever look at it except Ingrid — so I had literally no fear, no idea of what people would think, no desire at all, to impress anyone. I only wanted her to like it.

AN INTERIOR VAULT, 1976

This is one end of a small room which I built myself in part of our Mexicali project. It is not the most beautiful room, and not the most beautiful vault . . . but I include it because it is completely ordinary. It is simple.

There is nothing but the full-bellied curve of the vault, the white wall, the brown beam . . . the simple window, the crossed lattice strips of the ceiling.

I feel confident that if someone saw this picture, they would not know that it was built by an architect, but would assume that it was an ordinary cottage, in one of the thousands of forgotten villages, in Greece, or Mexico, or North Africa . . . and to have achieved that is very lucky, a beautiful accident . . . the fruit of twenty years of work.

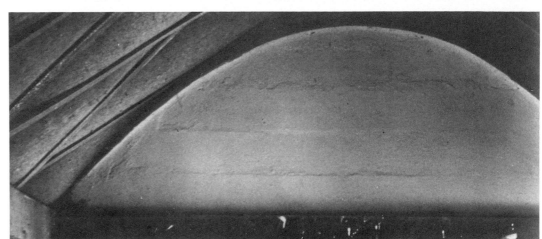

BLACK AND WHITE TILE

This is one of a series of about fifty tiles which I have made: each one different in colours, glazes, and design. Some are more exciting, some are more vivid in colour, some much more bold in their shapes. However, of all of them, this is the one which perhaps comes closest to the innocent mirror of the self quality which, I hope for, but which is so hard to find.

I believe some aspects of the human soul, in its more peaceful moments, look something like this.

Of course, no thought went into it at all. It took very much concentration, but about five minutes work.

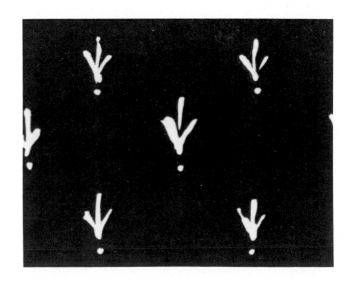

THE BERRYESSA HOUSE, 1982

This house is under construction as I write. We put the foundations in a few months ago. The client has now found the money to go ahead with the main structure itself, and at this moment, columns and beams and ceilings and stairs are all standing.

The building is almost like a miniature palace ... seven small buildings, dotted on a steep hillside, among white oak trees, each one with enough space, outdoor terraces, stairs, and connections to make the whole complex almost magnificent in feeling. The buildings include a tiny library, with a high coffered ceiling, a three-storey tower, with a meditation chamber on the top floor in a room surrounded by window seats.

All the buildings, except the out-buildings, are built in the combination timber-concrete construction which we have recently developed. The vertical loads are taken by the post and beam system, where columns and beams are simply pinned together by invisible steel pins. All shear forces in the building are taken by a one and a half inch concrete shell outer wall, which acts as a shell, ties the whole building together, and provides the buildings with a solidity that is rarely found in pure wood construction.

The foundation contains dozens of tiny hand-painted blue diamond-shaped tiles, which were glued to the formboards before the foundations were poured, and now glow and spark and give a touch of colour that brings the long grey foundation walls to life.

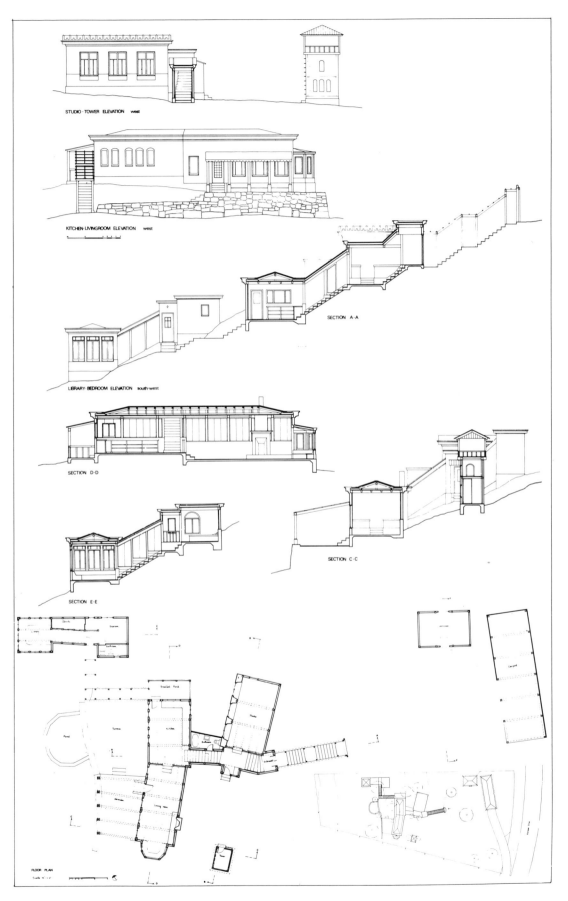

STUDIO-TOWER ELEVATION west

KITCHEN-LIVINGROOM ELEVATION west

LIBRARY-BEDROOM ELEVATION south-west

SECTION A-A

SECTION D-D

SECTION C-C

SECTION E-E

FLOOR PLAN

BAYS OF A TWO-STOREY BUILDING, 1970

This is another sketch I made, years ago, at about the same time that I first began to catch a glimpse of what I called 'the quality without a name'.

I made this sketch one afternoon, in the Centre, at a session where I asked everyone to draw, in a single simple image, the essence of what they thought we were trying to build: what a building made perfectly in the *Timeless Way of Building*, and *Pattern Language* could look like.

It was done years before we made any detailed studies of building systems: and it is certainly crude. But even now, years after, it does, in its very roughness, catch some of that fire in the grass, of the evening sun, the building made at a time of perfect unconsciousness . . .

CONSTRUCTION YARD, 1977

This picture shows one of our construction yards, during the summer of 1977, when we undertook a number of experiments to see how we could adapt modern uses of concrete, to be compatible with the kind of easiness, and relaxedness, which is described in these sketches . . . what you see, are samples of columns, beams, walls, arches, parapets, vaults, ceilings, roofs . . . and in all of them, we were trying to understand how to turn the use of concrete away from the high-industry, mechanical character which it has today, into a human and joyful use, capable of letting us build with the character these sketches have . . .

SKETCH FOR A CONSTRUCTION SYSTEM, 1977

Done at the time of a long series of experiments in concrete construction, this is another sketch, made in a few moments, to try and define, intuitively, the most important aspects of a building system. Rough as it is, it has a smoothness which I like very much. If I could build a building, in concrete materials, which had this quality, I would be very satisfied.

It is very orderly: not much funkiness, no strange shapes. It has immense windows, full of light; texture in the windows, in the roof surface, small ornaments at key places: yet all very rough.

Perhaps the great and beautiful Lansdown Crescent at Bath comes closest in practice to realizing what this drawing contains.

The boundary between the floors is marked, and ornamented. The door is very pleasant, marked by the half-circle light above. Even the ridge of the roof is marked by punctuating ornaments. The building is very strongly ornamented: and yet it is simple and smooth, almost severe – the ornament has not become ornate or heavy at all.

MODESTO CLINIC, 1972

The process of laying out this building is described, rather fully, in the *Timeless Way of Building*, Chapter 23. As I explained there, the actual process of construction was not in our hands, and indeed, it was the result of this building which finally convinced me that it was necessary for us to take over the task of building itself – since there was no way of finding a harmonious way of building, so long as architect and builder were separated from one another.

The arcade is crudely made, and lacks many of the qualities I now know it ought to have. The columns are not thick enough; there is no cornice on the roof. The columns have no bases. They have no capitals. The brickwork of the floor is machine-like and impersonal.

In a nutshell, the whole thing is crude and machine-like. However, I include it, in spite of all that, just because, at least in its scale, it has the fact of trying to build an ordinary arcade, comparable to the arcades of traditional times. But I include it mainly as a failure, to show, by contrast with some of the other sketches in this book, an example of something I have made which clearly lacks the qualities which I am after.

BRACKET, 1978

This is the formwork for a concrete bracket made to support a bay window. Since the shape of the bracket is complex, I made this cheap form by gluing styrofoam together with rubber cement – and so I could adjust the styrofoam until the proportions were just right.

I include this example, to make clear something that is true in most of the cases described in this piece: the incredible effect which an inch or two can have on the success of a shape.

All my pencil studies show different possible combinations of the thickness of the different layers, different angles of flare, and so on. One – the one I finally chose – works better than any of the others. Why?

The answer lies in the shape of the space next to the bracket. If you look carefully at the space created next to the bracket, you see that it is only in the case I finally chose, that this space *also* has feeling, and a beautiful smoothness . . . while in the other examples, the shape next to the bracket either does not have any character of its own at all, or it is too jagged, too rough.

In almost every case, when we pay attention to the shape of the space next to what we are doing, as well as to the thing itself, success depends on an inch, even a quarter inch.

It is for this reason that pre-cast elements just will not work. The exact shape of the space next to something

depends on the interaction of all the elements present. To be successful, each thing must be made, exactly, according to the place where it is in the finished building and there is no way that pre-cast elements can succeed.

Only control over the last inch, the last quarter inch, in every particular case, and immense care, comparison of all the possibilities, intense study, looking at it, looking at it again, and looking at it again, only that will bring success at the last inch.

DOOR FRAME FOR A COTTAGE, 1973

This is the front door of the experimental cottage I built in Berkeley, and described in the *Timeless Way of Building*.

Of everything in the building, this door is perhaps at the same time, most rich, and also most lost to itself, most unconscious.

I made the decoration one Sunday afternoon when no one was there. I drew the 'S's on two boards in pencil, in five minutes . . . cut them with the jig-saw and nailed them up. I was a little afraid . . . it was so marked, I wondered if it would look too strange. . . . I filled the 'S's with plaster a few days later: and then built the door itself at the same time. The door is made simply by nailing and gluing one-inch planks together: it is the simplest construction imaginable.

DOME IN THE SCHOOL AT BAVRA, GUJARAT, 1962

I spent seven months in the tiny village of Bavra, in Gujarat, India, trying to understand the village way of life, and its relation to the buildings. While I was there, I helped the village people build a school, because I did not want to be there without helping in some way.

There was almost no money, so I asked the village potter to make several thousand cone-shaped tiles each one about a foot long and four to five inches in diameter, with a slight taper. These tiles could be set into each other, nose to tail, and a long line of them, could then be 'riffled' like a pack of cards, to form a curve. I used these curves to form arches, out of which the domes were made – and then plastered the outer surface. The lowest tiles in the arch were filled with mud, to give them greater compressive strength.

At least forty or fifty people from the village helped me build this building.

BALUSTRADE AND SEAT AT THE MARTINEZ HOUSE, 1978

This balustrade and seat, outside a small workshop on the Martinez site, are entirely made of gunnite – sprayed concrete. The balustrade was formed with styrofoam cut-outs, made in the form of arches and flowers, set against a backing of sheetrock – and then shot with concrete, and screeded off to a smooth surface. The seat was built over an armature of plywood boxes – now 'lost' in the concrete, with all the key elements, shapes of the seat, back, railing, etc., also formed by guides, and cut off after the concrete was shot.

The whole thing is painted pale blue and white, and is now a meeting place for all the local teenagers in the area.

A very simple technique, makes a permanent and well-built thing, of a quality and level of detail almost akin to ancient stonework – although much cruder, and of course cheaper.

HOUSE FOR BODEGA BAY, 1981

These two sketches are the rough pencil drawings I made after laying out this house in Bodega Bay, north of San Francisco, for the first time. The final design was modified, but the spirit of the house, the feeling of it, was already quite clear in this earliest drawing.

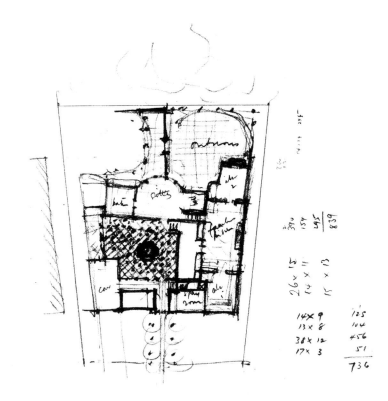

TEN-STOREY APARTMENT BUILDING FOR SAPPORO, JAPAN, 1982

This building, commissioned by Mrs Keiko Inoue, of Sapporo, is a simple apartment building, with shops on the two ground floors. Mrs Inoue and her family will occupy two houses on the ninth and tenth floors.

This building created an enormous challenge. As I have explained in *Pattern Language*, I have serious doubts about the wisdom of building any buildings – especially residential ones, which are more than four storeys high.

On the other hand, almost everywhere in the world, the development process and the high price of land are causing people to build apartment buildings which are six, eight, ten and twelve storeys high. It is happening on such a vast scale, that it seems absurd to turn one's back on it. I asked myself the question: if we have to build a high apartment building, can it be done in such a way that the apartments somehow feel personal?

In this building, each floor has an entirely different character. There is a morphological gradient, inside the building, caused by distance from the ground, closeness to light, and the decreasing floor area, which form a progression of different plans, on each of the ten floors. Also, no two apartments are alike. Each one has its own character, according to its position in the building. The internal structure of the apartments is very traditional, with a gradient of privacy, a light gallery towards the light, traditional mat rooms, and alcoves on the inside. The coherent concrete structure of the larger building co-exists with a second, entirely separate, and coherent wood structure for each apartment, within its space.

The legs are there, because the back part of the building is built over an existing hospital . . . and it was only this which made it possible to place enough volume on the site, within the existing regulations and local height restrictions.

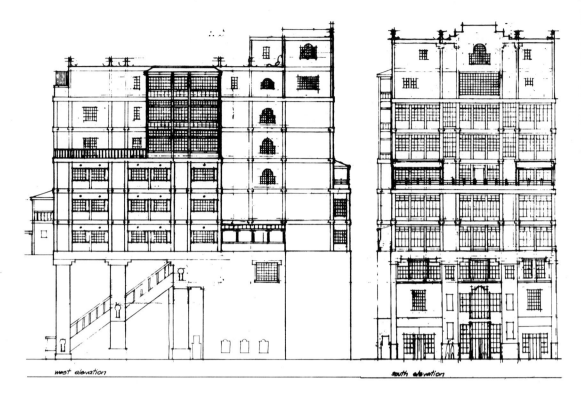

west elevation   south elevation

TILE FOR A FRIEZE, 1975

In the twentieth century, tiles and ornaments have become either casual (blotches of colour, glaze and clay), or far out (supergraphics, bold colours, op art). Simple classical designs – drawings which keep simplicity of shape and colour, but also come from a complexity that originates within the shape itself – are almost unknown.

This is a tile, with a version of a fleur-de-lis. The design arises simply from the impulse to make each part of the design positive – and yet to do it in a child's way, without exaggerated boldness, simply allowing the shape to become whatever it becomes, as one follows the dictates of the eye.

This tile, one of the first I ever made, I imagined as one of a series of similar ones, along the top of a wall, over a window...

EPILOGUE

Slowly, painfully, for the last few years I have been trying to make buildings, in which a little pure love, a little nature itself shines through, and which have none of that brooding, ego-filled artificiality which is almost all that we have learned in our own time.

In one way or another, the buildings of our time are works of prison architecture. Harsh or soft, large or small, concrete or wood, done by developers or done by architects, they do, with one voice, express the captured soul.

Even those made carefully as works of art, done with great care, photographed, and hung in museums, are still essentially at heart all works of oppression and imprisonment.

This has come about because the processes of building in our time — the system of production by which buildings are made — is a mechanism, a set of procedures which pass a building from person to person, which allow only rigid and mechanical things to pass through ... because there is no way a person can produce a building that is innocently sung direct from his own soul. I do not say that architects no longer try to do this ... but the process they are part of makes it quite impossible, for the most earnest and serious, as much as for the false and artificial.

In this sense, then, the works shown here, sketches, fragments ... are works of liberation, in which the process itself is altered, to permit the flight of the uncaptured soul.

In the *Timeless Way of Building* I have described that quality which lies at the heart of all things, the quality which has no name, but comes about when something is so pure, that it is free, like a whisper in the grass.

Of course, I have hardly been able to create this quality in its deepest sense at all so far. A few times, perhaps, by luck ... or concentration ... there are a few dozen things that I have drawn, or made, which have a little of this quality. And yet of course, I have made many more things, designed buildings, built buildings, which fall short of this quality entirely.

It is for this reason, that the few drawings and photographs which appear in this piece are fragments. I have rarely shown a building, in its entirety, even if I am pleased with it, but more often some corner which has a whisper of this quality in it ...

One day I hope I shall be given grace to make a whole building or small world which has this quality in its entirety. But until that day comes, there is no point in pretending, and it is far better to show the few things which truly have this character, modest as they are, and therefore to be clear, than to exalt things which do not merit it, and confuse the issue.

Christopher Alexander and Jan Johnson at the School at Bavra

Edward CULLINAN

NURSED BY EXPERIENCE

'All Arts were begot by Chance and Observation and nursed by use and experience and improved and perfected by Reason and Study.' **ALBERTI**

In 1954, when I was twenty-two, my father was given a 99-year lease on what remained of the Belle Tout Lighthouse on condition that he restored or repaired it. This small, symmetrical granite structure was built by one of the Stevensons between 1829 and 1833. It is 270 feet up on top of the edge of the South Downs where they were shorn off by the sea that formed the English Channel. So the land slopes downwards inland, away from the chalk cliff edge, and Stevenson used this topology to produce a two-level building. The squat light tower is on the higher ground close to the cliff edge, the house is behind it with its front door on the protected side, a floor below, opening on to paving in a walled garden terraced downwards inland. Ordered, compactly composed, symmetrical and split level; a solidly suitable blip on a sliced-off hilltop.

It was much tampered with over the years and was finally ruined during the second World War, being used for target practice from the landward side.

During the summers of 1955 and 1956, we took down the ruined granite walls of the house and found, and left aside, sufficient cut stone and detail to restore the tower

and the ground floor of the house to Stevenson's original form and I found that to build things oneself and with friends was a source of inspiration; partly for the simple physical pleasure it provides but also because it draws out the building process and gives one time to cogitate on the way that materials might join one another to enclose the spaces that surround life. Since then I have used the building of small buildings, taking a long time over them, usually at weekends, as an inspiration and a release from my other life as an architect who works for clients, deals with governments and builds to deadlines for certain sums of money through the agency of builders.

But, while I lived on the site and learned from him, it was Mr Viner, a skilled mason from Llewellyn's building firm, who rebuilt most of the lighthouse to designs that I drew during my final two years as a student at the Architectural Association. My tutors at the AA were Denys Lasdun, John Killick, Arthur Korn and Peter Smithson, and I look back on that period as a high time, a time when the neatness and trim of post-war 'contemporary' (as it was called to avoid the overcommitted MODERN label) architecture was supplanted by an attempt to re-understand the young traditions of modernism not as utilitarianism, neatness or simple reductivism, but as an art of defining spaces and interpenetrating them with one another both horizontally and vertically, relating them to one another using sticks and frames and planes and skins in abstract and asymmetrical disposition; using opaqueness and colour and transparency and translucence to make spaces and places to suit a human spirit

that is assumed to have no further desire for rigid and separated enclosures. And in September 1955, I cycled across France in leggy tribute to see Le Corbusier's chapel at Ronchamp, also a blip on a hilltop, the southernmost hilltop of the Vosges mountains in Eastern France. I was enthralled by the masterful quality of this building, and two years ago, when I saw it again, in a thick winter fog and melting snow, with my children, I found that I still was and they too.

Ronchamp is an abstract, asymmetrically balanced, partly hollow-sided object from outside and an abstract, asymmetrically balanced, lit space within.

If one starts with the great processional, occasionally used, enamelled, pivoting door on the south side, one finds on its west side the high tower with which this description will end, and on its east side the low, stable, wide end of the south wall which narrows its base and gains height as it moves eastward and is the means through which strong, joyful south light enters the interior, controlled in quality and colour by many glazed openings (1). That wall ends high, exactly vertical, thin (2). This height and thinness is heightened further by a slit (3) between it and the east wall which has statue, altar, choir lofts and pulpits on both its sides, inside and out, for small and large gatherings (4). In turn the east wall diminishes in height towards its right-angled corner junction with the north wall: which is a flat vertical plane until it turns 180° into the building, rises above and over itself to create a towered lighting device to a chapel within the curve (5); and stops to start again with another same but mirrored device of tower and chapel so that the outer sides of the two curves/towers/lighting devices/chapels create the main or usual entrance to the place within (6). From the second tower the enclosing wall follows a continuous, tense curve from north through west to south, dropping to a low point where the roof water spouts off (7), then rising again to end by again bending through 180° and rising above and over itself to create not only the largest of the three towers but the largest side chapel within and a west flank to signal the processional entrance on whose east side is the low wide part of the south wall where this description began (8). All these walls exist between the detached canopy of the roof and the floor which is level in part and sloping in part, in functional, palpable counterpoint to roof and walls. The roof overrides the vital south and east walls but is held within the other walls and the three towers; the floor is all within, contained. All the graceful details and pieces are scaled and ordered to create the most controlled, finite, abstract, multivalent, inventive, and calmly lit space that I know of; a masterwork by a master, towards the end of a life of architecture. Though it is one of many buildings, people, teachers and experiences that affected me then, it reminds one that building has occasionally reached towards the sublime and I would love to approach such a wholeness of composition, such a logical sequence, in my own work.

Ronchamp, Le Corbusier, 1950–54. The numbers on plan are refered to in the text.

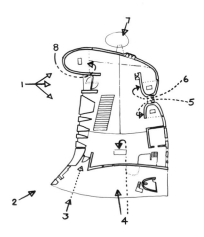

Of the two houses I made then, the first came after the first trip to America. It lies at the foot of a wooded bank on the edge of a small grassy patch through which a stream flows in a valley at Ashford Chase near Petersfield in Hampshire. The house faces south and is built on a concrete platform supported on ten posts which rest on the chalk substratum: the platform is long east–west and short north–south and serves to formalize the junction between woody slope and grassy patch. On to the platform the house is built: a continuous north wall, a plane of six-inch insulated blocks, tarred black for water-proofing; two south walls, two planes of the same blocks but painted white to make behind them a kitchen and small bunkrooms and bathrooms at each end of the platform with terraces (on the platform) in front; leaving, in the middle, a single large space for sitting, eating, meeting, talking, lying and, especially for my client and friends, to compose and to play music. The whole skinny structure is braced by two 'tilted up' concrete frames containing pivoting doors that open from the main space on to the terraces at either end. The platform was made by Concrete Ltd, the frames were made on the platform tilted off it by me, the walls and the roof structure were built by a seventy year-old gardener, Horace Knight, the roof by felt roofing contractors, the glass south wall and roof by Heywood Williams, and partly to ease this form of contract but mostly to obtain gracefulness from cheap materials, the construction was designed without inter-locks between trades, each material vigorously oversailing the next, mastering it. The long thin south-facing plan was to take advantage of our climate, the redivision of the spaces inside a small house into tiny spaces and a large one was to make a 'there there'[1] and there are long views, up to 55 feet through the house, all within a floor area of 600 square feet.

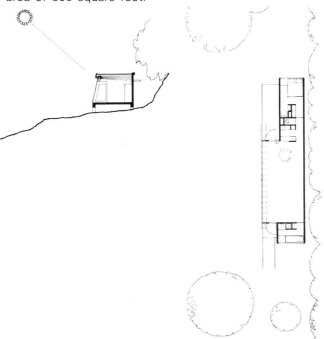

1 Gertrude Stein said of Oakland 'There is no there there' and I have often thought that the opposite is a good brief for a place.

The second house was built by me, my clients the Marvins, and a student. It is built beside a rock on a sloping meadow at the foot of the coastal range, over-looking the sea and Stinson Beach in Marin County, Northern California. It was for two people to live in and this helped lead to further investigation of the possibilities of changing the disposition of spaces in a small house. There are five glass-topped, concrete masonry towers with peephole windows which contain a dressing room for him, and a dressing room for her, with a double shower between, a guest shower and WC, a study with wine cellar below and a kitchen. These towers are placed beside and are the greenrooms for a long gallery made from sawn redwood; a gallery which contains a bed towards one end, a place to sit and meet in the middle where you enter, and a dining table towards the other end. The whole composition is placed at right angles to the hill and was to have been approached by a road straight up the hill, San Francisco style. A guest house and greenhouse built to the same pattern were to have been placed beside that road so as to exaggerate the 'natural hillside' quality of the meadow by an opposing geometry of road and building rather than lying with it; a lesson from the gridded streets of San Francisco that climb over the hills unshaken by the contours. The construction is toy-like.

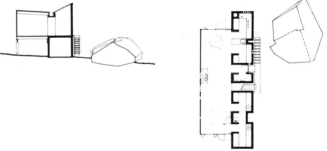

Opposite page *the House in Hampshire*
This page *the Marvin House in California*

33

From 1960 until the beginning of 1965, I worked with Lasdun on his schemes for student rooms at Christ's College, Cambridge and the University of East Anglia. It was very valuable; I learnt a great deal from him, he grew me up a bit. Whilst there, I designed three more houses, all in the context of bricky London streets, all between party walls with existing houses abutting them.

12 Bartholomew Villas replaces the bombed-off left half of the last of a row of nineteenth century two-house ('semi-detached') villas in Kentish Town. The design extends the rendered ground floor (or base in classical terminology) to contain garages and a flat; thickens the party wall to contain services, bathroom and kitchen in hollows therein; and in the cradle thus formed places a living sleeping gallery on the first floor and a music-making gallery on the second, both covered by a stepping lean-to roof, leaning to the crown of old chimney pots at the same 30° pitch as the remaining half of the villa.

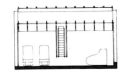

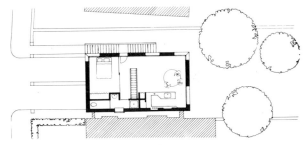

1a Greenholm Road is at the southern and downhill end of a lengthy and repetitive row of early twentieth century semis in Eltham in South London. The scheme adds a garage to the last one and its own roofs slope back up from this low point, tipped to allow southern sun into the long gallery above bedrooms and making a larger valley between the roofs of the end houses than is normal in the street.

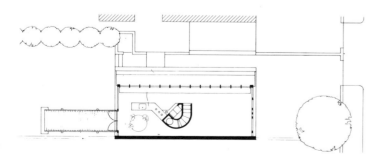

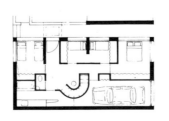

These two houses develop the theme of open gallery and small room and of simple expressive construction and, in the context of the city (London particularly), contain families. So does **62 Camden Mews,** the house that I built with my wife and friends in three years of weekends between 1962 and 1965. It is a small work but since it is small, it can be described in some detail, in an effort to describe the relationship between idea, construction and place.

This is a sectional perspective through the site, which is 25 feet wide and stretches 46 feet back from the street. The ground is drawn as existing, so are the two trees and the boundary wall; only the party wall to the next house, along the north side of the site, is new. It is 17 feet high and thickens twice from 9 inches at the top to 13 inches and to 18 inches to provide an upper ledge for future roof joists and a lower one for floor joists and to be more soundproof at the bottom next to the neighbour's garage and workshop. But since this is not the only way to achieve these ends, it also does it to look and feel stable and to be a visible, palpable support to the structure that follows.

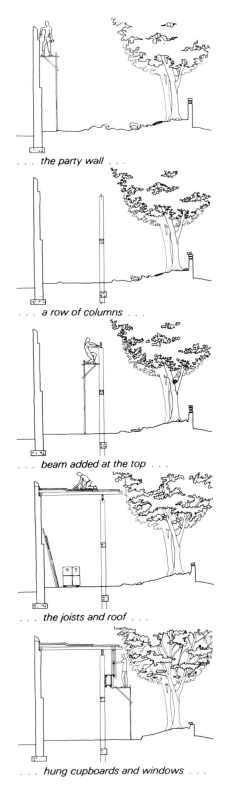

. . . the party wall . . .

. . . a row of columns . . .

. . . beam added at the top . . .

. . . the joists and roof . . .

. . . hung cupboards and windows . . .

This is the same with a row of concrete columns and a single concrete beam added, 9 feet 9 inches from the party wall to form the basis from which the south side of the house will develop. These columns and this beam conform to the London Building Act which requires all houses to be surrounded by an 8 inch incombustible wall; but more importantly they provide a solid, ordered and repetitive line of enclosure beyond, and against which the sticks and glass of the south wall can develop and over which the plane of the roof can rest supported.

A deep wooden beam is added to the top of the columns to support joists and to increase the sense of enclosure at the column line. It is fixed to the columns and the joists that rest over it are similarly fixed to the party wall at the ledge and at their outer ends and are notched to take a gutter; declared details at fixings and ends, a form of decoration. Over the joists is the roof which steps up twice towards the party wall, to drain forward and to increase the sense of protection below it.

In a 'do it yourself' house, it is good to have the roof on early in the job to protect materials and the building below for most of each week when no one is there.

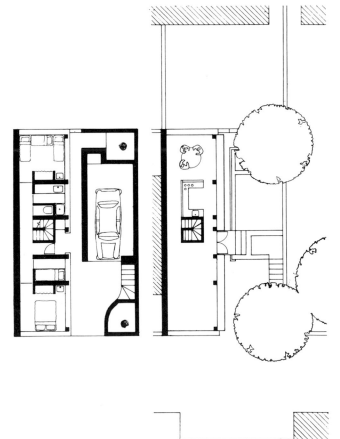

Over the columns the joists are paired and from those pairs hang two posts in tension from the bottoms of which hang the outer edge of a continuous cupboard fixed to the columns at its inner edge. Between the posts are uninterrupted sliding glass windows. The whole is thus an elaborate affair of sticks and glass hung out beyond the main enclosure; an outside that is inside that extends the actual width of the interior to 12 feet 6 inches and its sense beyond.

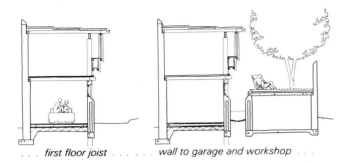

. . . *first floor joist* . . . . . . *wall to garage and workshop* . . .

The first-floor joists are supported across the concrete beam and a strip of window occupies the gap between floor and hung cupboard to extend the floor plane and to make the bottom outside corner of the site visible from the living gallery. Below, a brick wall encloses the fronts of the bedrooms on the line of the columns. The wall is 6 feet high to leave a view of sky and trees from those enclosed and separated rooms.

On the back half of the site this enclosing wall is mirrored by the wall to garage and workshop, solid roofed with terrace on top, the other party wall forming its southern side: a further development of the solid framework or cradle that contains the lightweight parts of stick and glass within.

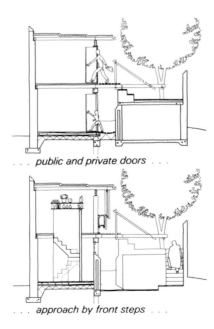

. . . *public and private doors* . . .

. . . *approach by front steps* . . .

Further forward on the site, roughly halfway down the house, the south side of the upper floor cuts back to place a front or public door on the column line, with a back or private door below it on the bedroom floor. Two floor joists project out to form a receiving platform in the porch and to carry handrails and then to be them. In response the garage workshop carries solid approach steps to just below those joist/rails: the solid framework rising to meet the lightweight insert, all to make an entrance.

The approach to the door is reached directly by front steps that rise from the front court or more elaborately by ramp that travels the length of the site in the crack between house and workshop. Internal connection is by a tiny stair that brings with it into the middle of the upstairs gallery the solid brick divisions of the lower floor, crowned by bookshelves, mantelpiece-like.

The two party walls and the ground are the established boundaries of the house which partly fills this territory or cradle with walls, sticks, skins and glass to make an interpenetrating indoor/outdoor world, perceived in many ways, that one can move through from boundary to boundary. The simple, quite cheap, building construction is used to reify the solidity of wall, the stickness of sticks, the invisibility of glass and the plane of the roof: it is not used only for the sake of simplicity itself.

From 1965 to 1969 I taught at Cambridge and designed and supervised the building of the Centre for Advanced Study in Developmental Sciences, a residential conference centre at Minster Lovell Mill in Oxfordshire, with Julian Bicknell and Julyan Wickham, Denys Lasdun having introduced me to the client. Minster Lovell village is a single street that runs along the north slope of the valley of the River Windrush from the old priory at the east end to the Miller's house and the pub at the west. It is on the edge of the Cotswolds and it has that particular combination of oldness, mild variety of building yet homogeneity of materials, of sandstone walls and sandstone roofs and a shared cross-sectional relationship of house to verge to street to verge to house that produces a sense of calm without challenge, that makes us fond of the Cotswolds. The Miller's house had been a private house since the 'thirties and stretching to the west a high drystone wall forms the north boundary of its landscaped gardens, which also contain a malthouse and a large barn, and the base of the mill itself that crosses the river. The conference centre is seen as a system for connecting and inhabiting and enlarging these buildings and for extending the village westwards. The connecting buildings have simply jointed timber roof structures that are supported on paired posts down on to concrete masonry walls. The cross-sections are fully occupied; wide for big rooms, narrower for smaller ones, the smaller sections being generally used to connect the larger ones. There are also glass-sided rooms and rooms with grass roofs where the contours of the landscape or the situation demand or permit them. The buildings are clothed in second-hand stone and attempt to occupy space in a manner that is sympathetic to the surroundings. There are no so-called 'Cotswold features' to be found.

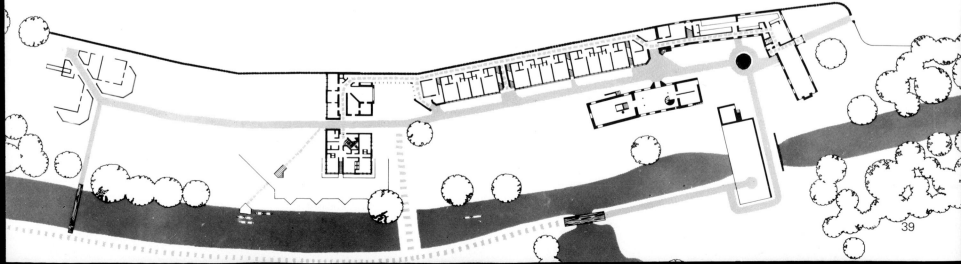

Between 1969 and 1973, much of my work was for British Olivetti Ltd and included the pleasure of working for clients, Carlo Alhadeff in particular, who had no doubt that architecture was an art – an essential art – and who were extremely keen on building buildings. Although building design undoubtedly has social aims and social consequences, as well as scientific environmental and structural engineering aspects and the design process is slightly amenable to systematizing, there cannot be much doubt that buildings require their own formal quality, their own ordering of place and object, their own deliberate harmonies or disjunctions through the manipulation of form and hollow as seen in light: architecture is active composition and can produce responses of pleasure or not as can music, painting and other arts. Sometimes it seems, this is doubted.

I designed for Olivetti, with Julian Bicknell, Michael Chassay, Giles Oliver and Julyan Wickham, the conversion of a large late-nineteenth century house at Haslemere into the residential part of their training centre next to the teaching building by Jim Stirling who had originally introduced me to Olivetti. Further conversions for their branch offices were of a Salvation Army meeting hall in Hove, a small 'forties factory in Thornton Heath, an asbestos warehouse in Cardiff, the 'twenties sub-Art Deco Ferrodo Brake Works in Bootle and the lower two floors of a 'sixties tower block in Coventry, and we designed them four new branches in Dundee, Carlisle, Belfast and Derby.

The site in Belfast is among suburban houses; at Dundee an industrial estate, although high on a hillside overlooking the city and the River Tay and (to our client's delight) the National Cash Register factory; at Carlisle on

the site of the old marshalling yard between a petrol station and a timber yard next to the main line from London to Glasgow; and at Derby on an extremely dull, flat industrial estate, then empty, now half-full, between a disused railway and half-drained canal. Had the programme of building continued, there would have been other types of site but probably mostly, like these first four, on unprepossessing industrial estates. So the buildings aim to stake out and to make their own territory and to have their own character, quite distinct from the other buildings on the estate. We visited many Olivetti branches and discovered a great need for flexibility between departments inside the building and for the possibility of future expansion; but since we wanted the buildings to seem whole from the start, we devised them as open-sided courtyards, complete along the street side and returning up the sides of the site as far as the initial accommodation allowed, leaving two temporary ends from which to expand towards completing the courtyard in future. The first floor is the main floor of these U-shaped buildings; it is both the *piano nobile* of Classicism and the working attic of the industrial tradition. It has a 4m wide space round its outer, longest perimeter, that is used with or without divisions as offices for salesmen, secretaries, managers and mechanics and for showrooms and workshops; inside this outer ring is a further 2m of storage and doorways to the offices and inside that 2m of passage, glass-sided, cloister-like, overlooking the building's own partial courtyard. About 1m down from the main floor and filling one of the two inner angles of the 'U' is a space for workers and visitors to meet in, to sit in, to have coffee or lunch in. This place is also the foyer into which you arrive from the front door via the hall on the service floor below. It combines occupants with visitors like the living room of a

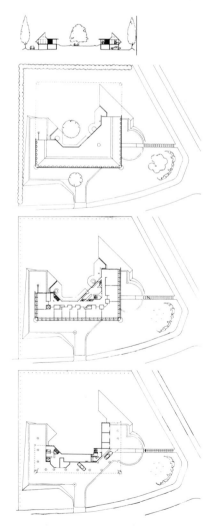

Olivetti Branch Offices, Dundee

house, and like a living room, it leads to, and from, more private places and out into the courtyard through French doors. The main floor in every branch is made from the same pre-made, dry-constructed, bits and pieces; the perimeter working space is spanned every 4m by plywood webbed trusses that rest on plywood fins along the outer perimeter and on to a steel frame that forms the roof to the 2m of storage and the 2m of passage along the inner perimeter. The plywood trusses and fins make a high section to emphasize and make enjoyable the working area, to let light in from both sides which avoids glare and sun from whichever side it shines. The steel frame over the inner perimeter areas, which is exposed and painted the colours of the local football team, makes a low flat ceiling over those areas, although it is high over the foyer where the floor drops away. The main floors are supported by a lower floor of traditional wet construction which contains stores, service rooms, main entrance and back door and provides covered parking round its perimeter; this floor mediates between the various special demands of the site, such as planners' demands for particular points of service access, car access and people access, slopes and perimeter road layout and the need for a main floor of regular layout, hence the various and irregular ground-floor plans.

Since he describes our aims in the result better than I can, I would like to quote Mark Girouard on these buildings.[2]

'They have none of the smooth glossiness of the conventional prestige office. They are made up of relatively cheap elements fixed together with no very great attention to getting exquisite junctions... This casual and economical approach to the finishes combines with the quality and quantity of the space and the gaiety of the colour to give the buildings their uniquely individual character.'

And of the working spaces in particular:

'The combination of the height of the roof ridge with the sloping roof, the barrier of the fins and the modest scale of the window bays makes the room both spacious and intimate; the service ducts in the window bays are just at the right level for secretaries to sit on and chat, or smart young salesmen to sit on and have dashing telephone conversations (sic). With glazing along two sides, through ventilation is possible; there is no glare and, whatever their orientation, all rooms get some sun at some stage. This seldom gets beyond a corner of a room or the ledge beneath the triangular windows — for working purposes enough to enliven but very seldom to annoy. The combination of roof slopes, fins and different glazing means that light is reflected with different and varying intensities on differing surfaces and prisms; all rooms are suffused with varied and living daylight and, even on a dark day, there is seldom any need for artificial light. The outside world is divided by the lower windows into a series of separated vignettes, each seen through the triangular frame of their upper portion and the horizontal gutter outside; it is an effect which makes any landscape seem curiously exciting, not only the obvious ones like the panorama of Dundee seen from the Dundee building, but the little slices of railway, sheds, piles of timber and old bricks, derelict ground or near villas of Carlisle, Derby and Belfast.'

2   *The Architects' Journal* June 27, 1973.

During 1970 the office was re-organized into a co-operative or 100 per cent partnership. Under this arrangement no member, architect or otherwise, is paid a wage: instead they receive a percentage of each fee bill when it (or a group of small ones) comes in. At that time the percentages were kept within the range of two to one in line with the suggestions of the Industrial Common Ownership Movement, but recently they have gone to three to one (as at Mondragon in Spain) due to increasing differences in age between members. When new members join us, there is a six-month period during which they try us and we try them, and if both are agreed, we then re-adjust everyone's percentage to include them; this by each person suggesting a percentage for themselves and for everyone else, followed by a discussion aimed at agreement. It is by no means perfect, it certainly does not produce instant peace and calm in a room full of egos, but it is aimed at sharing responsibilities and rewards, allowing egos to exist and to contribute what they are best at and effecting a measure of social progress where one has the power to do so. It may seem slightly funny to talk of social progress at a time (1983) of economic despondency and conservatism, when the mood among architects is to doubt or be uninterested in the social aspirations of the modern movement: but I am sure that since architects think of and imagine futures in the form of new buildings and groups and places, it must be hard for them to avoid thinking of a kinder and fairer society to occupy those buildings and in turn to stimulate their imaginations. In an introduction to her books, Ursula le Guin, who writes of futures with an optimism concerning the human spirit that must be unique in science fiction, holding that the pursuit of art, by artist or audience, is the pursuit of liberty, quotes Freud's opposing view that 'the artist's work is motivated by the desire to achieve honour, power, riches, fame and the love of women'. We ride between that and such statements as Sullivan's 'Democracy depends on its Architects as much as on its statesmen. Until Democracy produces a good architecture it cannot produce a good life for its citizens.' 'Almost everything', said Thomas de Quincey 'has either a moral handle or an aesthetic handle', and Architecture needs a grasp on both.

The contrasting and interrelated aspects of optimism and pessimism are now more sharply raised by the design of houses than by any other aspect of architecture I know of. During the 'seventies, we designed five schemes for Local Authorities, two for a housing association and three for private clients. Of the three private developments, two were sold on, after we had obtained planning permission, and were built in a destructively modified form, and one did not add up financially; so this description concerns only schemes built for Local Authorities and a housing association.

Local Authorities and public corporations have given architects nearly all their commissions for groups of houses, flats and maisonettes from small to vast. There have been failures and partial failures: the Local Authority house or flat seen as a styleless bureaucratic non-event; the Local Authority house seen as an opportunity for large-scale formal muscle-flexing at the expense of domesticity; and recently the house for ordinary people seen as a need consciously to design ordinary houses in the pseudo-vernacular. The dilemma in general is that Local Authorities build for low budgets, although lastingly they hope, for the poorer half of the community, whom they generally and wrongly assume to be the less adventurous, at least as far as the aesthetics of lifestyle are concerned, while their architects want to imagine and to compose futures. Nevertheless the partnership of Local Authority and architect has often made good places and has sometimes made the most significant changes and advances.

We can for example see our received notion of suburban house and suburban place deriving from the period when the Arts and Crafts movement developed at Port Sunlight and Letchworth, and the pre-First War LCC; our re-understanding of the town house and the maisonette from the LCC of immediately post-Second War and schemes like Roehampton, and the possibility of making satisfactory ways to live above one another at highish densities in the recent development of stepped and interlocking sections, hanging gardens, Alexandra Road.

I will describe our own work for Local Authorities and a housing association under six headings to give an idea of the location and the aims of the schemes:
1 House and garden; community suburb;
2 An old suburb intensified;
3 An old street corner;
4 Forecourt, building and garden;
5 A centrepiece;
6 Downtown new city.

1 HOUSE AND GARDEN: COMMUNITY SUBURB

At Highgrove in Ruislip for the London Borough of Hillingdon, we designed and built 113 houses and flats between 1972 and 1977. The density of 140 people to the hectare is a very common one among Local Authorities building in suburbs, although it is considerably higher than is normal in private semi-detached developments. This used commonly to result in schemes of narrow frontage (usually 3.6m) houses, more suited to higher densities, in blocks of six or eight floating about in a sea of municipally maintained grass. Instead, we invented, as a development of my early houses, 9m wide houses joined to others in groups of four, so that each house's wide frontage is able to overlook and to possess a large hedged-in garden, suburban style. Two roads run parallel to the houses; the front doors open off protected paths or mews at right angles to the roads and the whole is arranged as a planted, avenued, and hedged formal garden above which sail the planar roofs of the stepped-section houses. My co-designers were Michael Chassay, Mark Beedle and Brendan Woods.

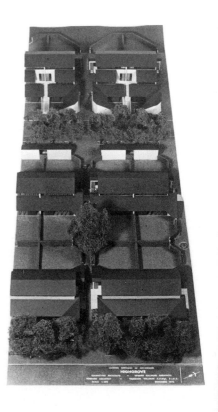

Highgrove, Ruislip

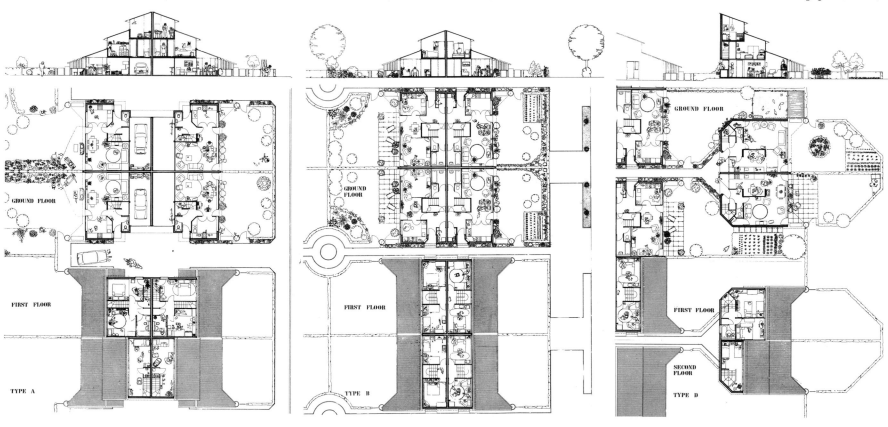

GROUND FLOOR

FIRST FLOOR

TYPE A

GROUND FLOOR

FIRST FLOOR

TYPE B

GROUND FLOOR

FIRST FLOOR

SECOND FLOOR

TYPE D

Green Lane   Site road   Chester road

2 AN OLD SUBURB INTENSIFIED

In the first half of the twentieth century, the expansion of London to the north-west spread detached and semi-detached villas far up the sides of the Northwood Hills; in the second half, these villas are being replaced by denser development. Our scheme (also for the London Borough of Hillingdon), instead of replacing houses, converts some of the villas into apartments and leaves others, while using part of the land between them for new houses. At the same time it sets a pattern for further development as further land is obtained. It does so by combining south-facing terrace houses with east- and west-facing villas. The single new road is located by the villas and a matrix of pavements leads through the site and past all the front doors.

My co-designer was Anthony Peake.

Chester Green, Northwood

3 AN OLD STREET CORNER

Between 1974 and 1976 we made for the Solon Housing Association a small building containing four two-storey houses that forms the end of a row of late Victorian double-fronted villas at Selhurst Road in Croydon. It is a simple box with a detached lid or roof plane on top, from which hangs the south-facing balcony for the upper two houses. The upper two houses are reached by a stair/path which is a development from the rendered stair on the villas next door; an anchor and the last beat of a rhythm that starts far up the street. My co-designer was Anthony Peake.

Selhurst Road, Croydon

45

4 FORECOURT, BUILDING AND GARDEN

Between 1974 and 1979 we designed also for Solon a building in Westmoreland Road, Bromley, that is made up from six four-bedroom, eight three-bedroom, ten two-bedroom and twelve single-bedroom houses and flats; their front doors open on to the forecourt on the road side and their gardens, indented balconies and terraces, face south across a large shared garden. The forecourt is large scale and formal and its grouped openings and doors open on to three pavements (at ground level, one floor up on a pergola, and higher as a gallery) which stepping back, overlook one another and the forecourt. The pavements connect to ramps which connect to Westmoreland Road and enclose and define three sides of the forecourt — grandly. On the garden side each apartment is more separately expressed and decorated with planting troughs and climbing wires, rails, trellises, steps, recessed and hanging balconies, French windows, terraces, gates, gardens and lighting and power points; the paraphernalia of house life at the edge of the garden. My co-designers were Brendan Woods and Sunand Prasad.

Westmoreland Road, Bromley
Above *forecourt and roadside*
Left *garden side*

Leighton Crescent, Camden

5 A CENTREPIECE

Leighton Crescent is a mid-Victorian crescent in North London and between 1977 and 1980 we made four three-bedroom houses and twelve flats, for the London Borough of Camden, on the site through which the axis of the crescent passes. The houses and the flats above them are quadripartite in plan, as at Highgrove; each has a front door to the central glass-roofed stair and each has three full-height French windows along its frontage. This gallery plan in the flats (kitchen, living, bedroom side by side along the frontage) allows the tenants to choose very open, somewhat open or closed plans by moving one or two partition walls. Although the flats are of today's small scale, the widely spaced, large French windows and the pattern of doorguards, rails and support poles outside them and the 'cornice' of the windows of the top flats, with the major axial approach ramp, give the building a size and a scale to make a fitting new centrepiece to an old Victorian crescent. My co-designers were Philip Tabor, Mark Beedle and Michael Chassay.

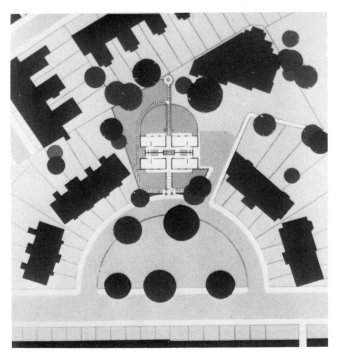

6 DOWNTOWN NEW CITY

Between 1978 and 1982, we made 158 new houses and flats and two shops on the outer edge of the central area of the new city of Milton Keynes. The central area is laid out on a grid of wide plane-tree-lined boulevards with secondary roads at right angles to them and we willingly accepted this format since it is clear that it will one day turn the various schemes of small terrace houses into a coherent place, its scale enlarged by the avenues of trees. Our scheme uses wide houses and flats with large gardens that open on to protected, shared open spaces that are deliberately and formally laid out as useful landscape: a children's figure of eight cycletrack, a formal rose garden, a castle on its ramparts, lanes for walking, small children's playspaces and one for larger children, an herbaceous hill, places for ball games and two tree-lined apses at the bottom end. The houses are high and gabled to the boulevard, they step down the hill by stepping up first to exaggerate the hill, they interlock three floors with one to make corners, they make a valley of the cross road and rise to three floors to make gateways to the mews and central path and for the elderly they make a court. The seventeen different types of house are used everywhere as a response to and in turn to enhance the situation they are in. Like the rest of the central city, they are made of light brick but are combined with vivid colour to underline further each situation and to produce lightness and brightness in the scheme. My co-designers were Anthony Peake, Giles Oliver, Michael Chassay and Sunand Prasad.

Milton Keynes, housing and shops, below *site plan* right *details and elevations*

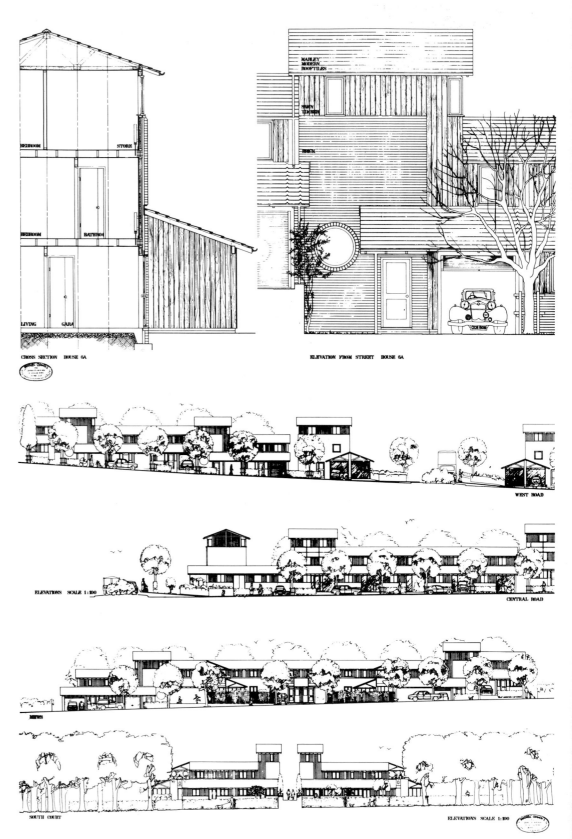

That is the history; there follows the present.

At High Wycombe, a late nineteenth century house, partly demolished and then extended to make a conference centre.
Above *sectional perspective through the east wing*
Below *plan of the site*

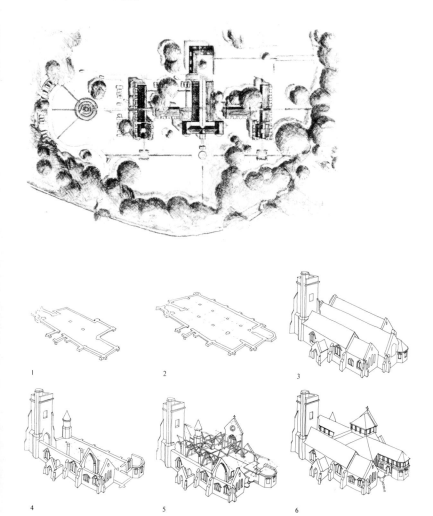

At Westoning Manor new buildings for mentally handicapped people complete a courtyard, growing from one existing side. Workshops, class-rooms and meeting rooms form the sides, houses make the corners.

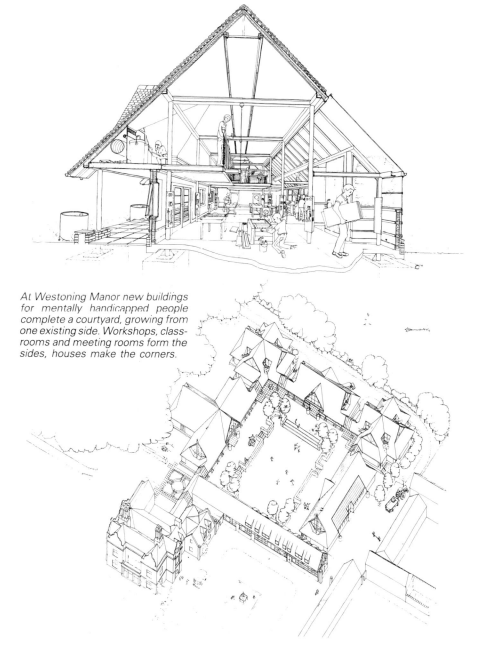

Barnes Parish Church is a small medieval building, much altered, much added to over the centuries and very severely damaged by fire in 1978. We restore the medieval church and add a new nave to the north which creates its own crossing, lantern and transepts; the transepts terminating in a fragment of an early twentieth century vestry at the east and an old turret at the west. The new is to connect, to respect and to glorify the old.

Giancarlo De CARLO

THE UNIVERSITY CENTRE, URBINO

1 THE TOWN AND THE UNIVERSITY

I designed the first student residences for the University of Urbino in 1965. At that date, a few of the most ancient Italian universities — such as Pisa, Padua, Pavia — did have student colleges, but they had been founded centuries earlier and had always been reserved for a small number of selected students. Between the wars some 'Case dello Studente' were built, for students who had no particular academic or social claims: these were rather like third-rate lodging-houses, with restaurant plus recreation rooms on the ground floor and shared bedrooms above.

In the late 'thirties, the government 'gave' to Urbino one of these Case dello Studente, which was clumsily stuck in the heart of the ancient town: it took up a great deal of space and yet contained only a few dreary and awkwardly shaped rooms. Students at the University would have totalled no more than about 300 at that time, and the ones from outside Urbino preferred to take a rented room, although these were even less convenient.

After the Second World War, the University began to grow rapidly, until undergraduates totalled almost 10,000 by the end of the 'seventies. Then the rate of growth steadied and now enrolments are just under 12,000. A third of this number actually attend courses regularly, while the rest only show up here (coming from a wide area of central–southern Italy) at exam times.

This fact, in itself undesirable, is actually lucky for Urbino, which is only a small place despite its fame. Its 7,000 inhabitants constitute a social structure which would be too frail to cope with such a huge influx permanently.

When students surge into the town three or four times annually and are added to the ones living there all through the academic year, the strains on the city begin to show. The cost of living has climbed to the levels of the biggest Italian cities. The suburbs continue to sprawl outwards because of the increasing demand for accommodation, although the resident population of the town is static. The strain of all this is most deeply felt in the ancient town centre, overwhelmed by motor traffic and drained of the less affluent social classes, with skilled crafts and trades being replaced by ephemeral commercial activities, while the ancient buildings are converted behind their facades to provide rented accommodation, and there is congestion in the public spaces which once provided somewhere that townspeople could meet and relax, now thronged and hectic with students and tourists.

Of course there is also a brighter side to the problem. The city and its countryside that used to be very poor for most of the last hundred years, now enjoy a certain prosperity; the society which had been stagnating in the absence of open contradictions has been galvanized by a wave of energy, somewhat brutal but stimulating in its way.

URBINO HISTORIC CENTRE

COLLEGIO DEL TRIDENTE (B)

COLLEGIO DEL COLLE (D)

CONVENTO DEI CAPPUCCINI (E)

COLLEGIO DELL'AQUILONE (C)

COLLEGIO DELLA VELA (A)

SERPENTINES

2 THE PROGRAMMES OF THE UNIVERSITY AND THE TOWN

The University of Urbino is under the guidance of an outstanding person, the writer and literary critic Carlo Bo, who has been its Dean continuously for more than thirty years. He has made almost as deep an impression on the town in these years as Federigo di Montefeltro did in the fifteenth century. Among his many achievements is that of having immediately understood that the University's rapid expansion was going to have important repercussions on the town and so a programme would have to be planned to cushion its detrimental effects and encourage the beneficial ones. The programme was based on two propositions: expand the teaching and research facilities in the ancient town centre and move the student residences to new buildings outside the city. Besides strengthening the links between townspeople and the university population, the development in the town centre would make it possible to retrieve a number of large buildings, constructed between the fifteenth and seventeenth centuries, that had fallen into disuse and would have been difficult to convert to small-scale activities without destroying the integrity of their architecture. The decentralization of student residences, apart from taking the pressure off the overcrowded centre, would provide the students with accommodations more suitable than rented rooms or the existing 'Case dello Studente'.

These two points appeared so obviously sensible that they won the full support of the townspeople and municipal administration, and were included in the town's Master Plan, then being drawn up.

It so happened that I worked on both the University Programme and the town Master Plan, first in the planning stage and then in their implementation. For the town, I realized some of the many designs I had worked out, including (in the historic centre) the underground garage below the Piano del Mercatale, the restoration of the Rampa by Francesco di Giorgio Martini and the rehabilitation of the neo-Classical theatre, which was built on top of the Rampa in the nineteenth century, effacing its memory for more than 150 years. Working for the University, in the southwest section of the old town centre I housed the Law Faculty and Magistero (School of Education) in the remains of two convents, built in the fifteenth and sixteenth centuries and abandoned by their religious Orders in the nineteenth. Outside the town, on the hill known as the Colle dei Cappuccini, I built in 1965 the first part and between 1975 and 1980 the second part of the system of student residences. Through the work I did for the University and the town, I was able to observe from two opposed standpoints the contradictions emerging in their relationship. At the same time, because of the lack of economic resources and the difficulties that can crop up when one's work covers a long span of time – nearly thirty years – in a place of outstanding historical value, I was able to reflect on the consequences of each phase completed and so derive certain pointers for the subsequent phases. After a while, a third proposition came to be added to the two original ones: this was that the University should not merely draw energies from the town but should also contribute an energy of its own to the town. The public spaces of the University, which are numerous, must be open to use by the townspeople, just as the public areas of the city are available to the students.

3 THE COLLEGIO DEL COLLE

In the sysem of student residences in Urbino, the section built in 1965 is now called the Collegio del Colle. It is reached along an avenue of cypresses, centuries-old, which once led to a belvedere where the monks from the Capuchin monastery used to gaze contemplatively over the valley. The monks have gone now and the monastery has become a home for the aged;

in place of the belvedere there is now the entrance to the central part of the Collegio, developed on three cylindrical bodies that intersect and are vertically staggered in relation to each other so as to follow the contour of the sloping site. In the central section, containing reception, lounges, restaurant, kitchen, library and, near the entry, a conference room, there is a network of footpaths which run against the contours and so, apart from a few flat stretches, are mainly made up of stairs and ramps. The paths lead to the student lodgings which are set along the arcs of circles following the contours and are hence on a flat level except at the points where the hillside slopes more sharply, in which case the body of the building is made to follow it by being stepped up or down.

Each segment of the arcs is set on two levels and each level of a segment is divided into three parts: the first, facing outwards on to the valley, comprises the bedroom, always single; the second, in the middle, comprises the entrance and bathroom; the third, facing inwards on to the hill, is level with a stretch of a covered walkway. The sequence of the various segments produces two different levels of walkways running along the whole length of the arcs and serving the student lodgings.

Adaptation to the steep contour of the land is secured by the horizontal and vertical sections of the segments, which are modular both in plan and elevation. In plan, the modules are not square but trapezoidal, their longer sides being splayed outwards at an angle of 6°, so that the sequence of segments is arranged in an arc and can slope down along the curve of the hillside. But where the curve is interrupted and there is a sharp change in level, the segment can be staggered horizontally or vertically or both for the equivalent of one or one and a half modules to follow the contours of the terrain.

This technique made excavation practically unnecessary, and, where it was unavoidable, the soil – pure clay – was utilized to make bricks needed for the construction. Brick is in fact the dominant material, together with the unfaced concrete used for the external horizontal structures, especially for projecting sections.

Twenty-five years after, now that the plants have matured and the materials mellowed through weathering and use, I feel that the Collegio del Colle is like one of the many 'seeds of the city' that one can still find in the ancient centre of Urbino: like the medieval nucleus surrounding the steps of San Giovanni or the nucleus opposite the Porta di San Bartolo.

I have been observing the Collegio carefully, through its brief existence, to see if it really had achieved its purpose and was merging into Urbino's scene, becoming an integrated part of it. As far as use is concerned, I can now see that its main fault lies in its too clear-cut separation of the individual life going on in the student rooms and the collective life in the central section on the hill-top, with no intermediate stages. As for its power of merging into Urbino's scene, I feel that the design had the premises needed to achieve this but that the building was too small in scale to generate enough energy to realize its potential. 150 people are too few and the 1,200 metres to the city centre too far to achieve complete blending with the town, despite all the care that had been devoted to grasping the arcane secrets of the place and embodying them in the design.

4 THE PARAMETERS OF FORMAL COMPLEXITY IN URBINO

I often say that Urbino is a town with a dual nature, because its complexity is woven out of a combination of opposites. It is small in terms of population

CONVENTO DEI CAPPUCCINI (E)

COLLEGIO DEL TRIDENTE (B)   COLLEGIO DEL COLLE (D)   COLLEGIO DELLA VELA (A)   URBINO HISTORIC CENTRE

COLLEGIO DELL'AQUILONE (C)

SERPENTINES

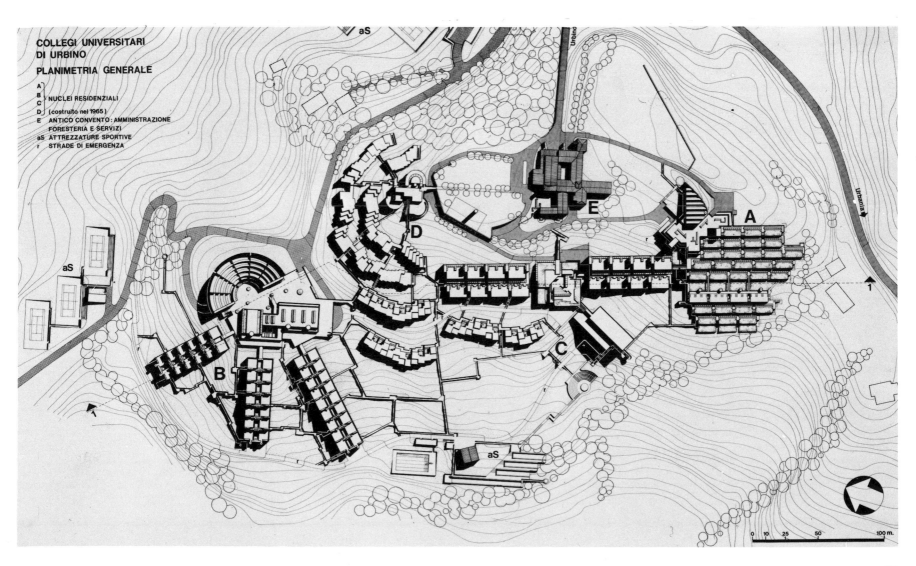

COLLEGI UNIVERSITARI
DI URBINO

PLANIMETRIA GENERALE

A
B   NUCLEI RESIDENZIALI
C
D   (costruito nel 1965)
E   ANTICO CONVENTO: AMMINISTRAZIONE
    FORESTERIA E SERVIZI
aS ATTREZZATURE SPORTIVE
r   STRADE DI EMERGENZA

but it strikes the newcomer as a great and glorious city; provincial but a capital; peripheral to the economic circuits but central to culture for the special activities that it performs; it is detached and self-absorbed yet at the same time enterprising and enquiring; it has survived intact in time but has also changed profoundly while responding to the needs, expectations and tastes of the various periods it has lived through. The dualism running through its existence is reflected in its environmental structure and this renders it complex and intricate.

If one wants to design in Urbino, I feel it is essential to enter into its complexity and discover the parameters controlling it. The parameters which I discovered after long observation are numerous but here I should like to pick out four which have probably had more influence than the rest on my work.

The first is that the pattern of the town and the pattern of the countryside are homologous. If you analyse a 'section of the cultivated countryside – the layout of the crops, the set of the furrows, the alignment of the ditches, the relation between the location of the trees and the slope of the land, you realize that nature here is man-made. If you analyse a section of the urban fabric – the way it follows the landform and stands out against the sky, the way it twists and turns on itself – you see that the man-made is natural. But what is really remarkable is that the man-made quality of nature and the natural quality of the man-made both obey the same aesthetic laws. This is why it is difficult from an aerial view of the city to separate the ancient town centre from the whole pattern of surfaces and lines by which it merges into the landscape.

The second parameter, correlated to the first, is that there is no marked boundary separating the ancient town from its natural environment. Although it has its circle of town walls, nature merges subtly with the city, forming islands along its walled perimeter and sending green spurs into the spaces between its

Collegio del Colle below *general layout,* right *facilities core*

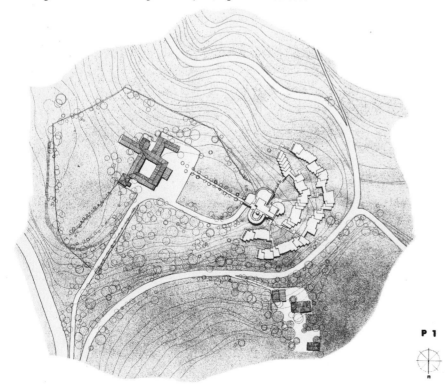

P 1

buildings. The town likewise merges with nature, through subtle reflections of it, as in the hanging gardens of the Palazzo Ducale, or by sending out offshoots towards the outside, like the church of San Bernardino.

The third parameter is the unity of the materials used throughout the ancient centre. The outer paving, walls of buildings and roofing are all in *cotto* (fired brickwork), used with infinite variations, according to period, economic value of the buildings, the culture and aesthetic taste of their builders. Sometimes the *cotto* is combined with stone, but these are specific cases where the co-existence of the two materials has a precise and virtually self-evident significance, so it does not result in over-emphasis or rhetorical distortions.

The fourth parameter, finally, is this: where the architecture faces inwards towards the town it is restrained and domestic in character, while where it looks outwards to the countryside it is made magnificent and glorious. Even the Palazzo Ducale, raising its turrets in a bold and mocking challenge to Rome, is restrained and modest on its other side, where it faces on to the square. In all periods, from the time of the Roman foundation to the neo-Classical transformation, this skilful contrasting of external and internal has been constant and it has given the town its irresistible and haunting fascination.

5   UNIVERSITY DEVELOPMENT AND THE NEW STUDENT RESIDENCES

The University in Urbino is 'free': which does not mean it is a private university, because it is a public one and follows the same criteria as the state-run universities with regard to student fees, syllabuses, appointment and salaries of teaching staff, and so forth. Its freedom consists of the fact that it is not financed by the state and so it can manage its slender supply of funds without reference to anyone else.

This may seem little enough, yet it is sufficient to secure a flexibility quite undreamt of in the state-run universities. It can dig its toes in against the bureaucratic proliferation so rampant elsewhere, it can organize special courses in the fields it is best qualified to handle, it can go to the banks when it needs backing for its development programme.

Ten years ago, the Dean – with a flash of imagination and counting on inflation – got a large loan from an important Lombard banking house and decided to invest it in construction of student residences which could accommodate 1,000 students in addition to those already living in the Collegio del Colle, making 1,200 in all. I must confess that, when I was consulted about the plan, my advice was to build various nuclei and distribute them at different points around the town, so that they would be merged with other urban activities. The idea of a specialized estate for 1,200 students, concentrated in one place, caused me some concern. But my suggestion could not be acted upon, because the University owned only the large site around the Colle dei Cappuccini and could not purchase any others. Besides this, the town council flatly rejected the idea of splitting up the plan and insisted that the site should be that laid down in the town's development plan for university expansion, while the bank preferred the mortgage to be secured against a single piece of property.

In this way I found myself having to design the new project for the same area where the Collegio del Colle had been built. Ten years earlier, when pictures of it began to appear in magazines, some commentators observed that it was an open-ended system and lent itself to indefinite expansion through the addition of further accommodation segments to those already there.

But ten years later, when I had to deal concretely with the issue, I immediately realized that this was really not the case. 'Open-ended' systems are still 'structures' and they can go on growing only up to the point where the system of relationships connecting the different parts with the whole and with one another begins to break down. Major growth over a short period puts a strain on any structure: the rate of change is too rapid to redress the imbalances created and re-establish an inner cohesion.

So one can affirm that an architectural structure is able to grow as long as growth does not jeopardize its cohesion: this cohesion being the result not of a static balance but rather a dynamic one achieved through variations and innovations that need time to settle into place.

The medieval 'seeds of the city' in Urbino's ancient centre have created fabrics which have taken on the forms of the various ages through which they have developed. Their remarkable coherence is the outcome of the myriads of variations and innovations that have occurred over the long span of time during which this development has been going on. In the case with which I was dealing, the development about to grow out of that particular 'seed of the city' embodied in the Collegio del Colle was going to be very rapid indeed, so there would be no time for the various imbalances to adjust themselves organically. The variations and innovations which organic adjustments could have produced would have to be introduced together and all at once. In other words, slow growth had to be simulated.

Simulation is an abstract process, as we all know, and we are familiar with the idiotic things it can lead to. The history of recent years is rife with counterfeits resulting from the assumption that architecture can fake any situation it likes, even if it fails to correspond to architecture's inherent reality. All the same, I feel that the simulation of slow growth is the only available antidote to counteract the inevitable degradation of the environment caused by rapid and unconsidered growth. Of course one has to be clearly aware of the limits involved, and try to avoid any distorting effects by keeping a firm grasp on the reality of the place where one is working. In designing the development of these student residences, the variations and innovations have been brought into harmony with the parameters that have always guided the town's formation and transformation; at the same time, they have been charged with energies capable of fostering the assimilation of the new project into the town. The parameters are those I set out earlier, while the energies are those that spring up out of the activities and situations that the University is capable of creating for the townsfolk, as well as for its own people, so that they may experience them and see them as new resources added to their city.

6   SIMULATION OF SLOW GROWTH

Playing on the metaphor of slow growth, I could say that the development of the student residences began with the Serpentines: the sinuous buildings that continue the configuration of the Collegio del Colle southwards.

On looking more carefully at these buildings, however, one sees that the similarity is only apparent. Instead of having two superimposed walkways on the side facing inwards on to the slope, they only have one; and each segment, instead of corresponding to one room, corresponds to part of a unit composed of a number of rooms, while the size of the segments is larger, even though the modular pattern is the same. When the first clay vases were made, in primitive times, their forms and decorations were still similar to those of the woven baskets used until then. They were different in weight, smell, feel, pliability, material; even the use and purpose of the artefact had

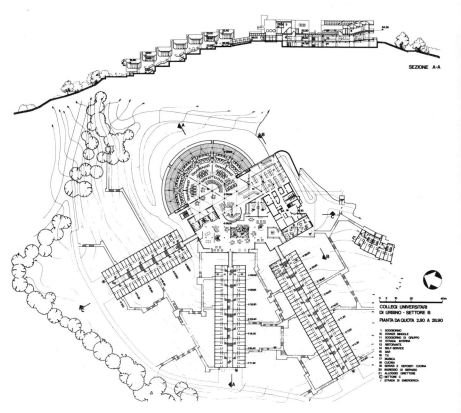

changed. But development came about slowly and, culturally, there was plenty of time for tradition to come to terms with innovation without strain.

Now it is not true at all that development of the student residences began with the Serpentines: in fact work was parallel in the three principal sectors into which it had been divided for practical reasons. But whether true or not is really unimportant, because the same approach, making use of an apparent similarity, was repeated in various different ways throughout. In the simulation of slow growth, there is, by definition, no difference between real time and simulated time.

7   INFORMATION ABOUT THE DESIGN

The design is fairly complex and it might be difficult to grasp all its features reading the few drawings that can be included in a publication. So I think it is worth providing some information about the quantities involved and the ways the various parts have been organized and related to one another. Even this may not be enough to explain everything, but then any part of a city which has grown up over a long span of time is likely to be too packed with images and events to be understood quickly and without effort.

The Collegio del Colle, built in 1965, has 150 rooms, a restaurant with kitchen and services, a sequence of lounges, reading rooms, recreation rooms (music, TV, games), and a conference room opening on to the main entrance hall.

With the expansion of the last ten years, three residences have been added to the Collegio del Colle: the Collegio del Tridente, with 352 beds, the Collegio della Vela with 222 beds, and the Collegio dell'Aquilone, with 408 beds, which also includes the Serpentines. The names may appear odd ('Trident', 'Sail', 'Kite'): they are derived from the half-joking nicknames invented by building workers on the sites.

Collegio del Tridente
Left *section and plans*
Above *dormitories*

Collegio del Tridente
Above *piazza. On the right are the lecture rooms
and the skylights above the restaurant*
Opposite page *restaurant*

Collegio del Tridente
Above *space for performances*
Far left *staircase leading to the restaurant*
Left *passage to the dormitories*

The Collegio del Tridente has single bedrooms, smaller than those in the Collegio del Colle but laid out so that eight of them are grouped around a lounge and a small kitchen. For every sixteen bedrooms – two sets of eight – there is a block of services comprising toilets, showers, wash-basins.

The collective amenities are more extensive than in the other residences because they are easier of access from the city and so better placed to foster links with the town's activities and university functions in the ancient centre. The piazza is the only one in the whole complex and has a series of premises for meetings and conferences, a restaurant capable of serving over 2,000 meals twice daily, a bar, lounges, recreation and reading rooms, a library and a newspaper and periodical library, and a theatre area for performances of various kinds.

In the Collegio della Vela there are mostly single bedrooms, but a few have two beds and are equipped with independent services and kitchenette. The single bedrooms are laid out in sets of six and each group has a small kitchen, bathroom facilities and a lounge, set on the other side of the internal walkway.

The collective amenities, in addition to the lounges, also include a restaurant and a big auditorium which can serve as cinema, theatre, conference room and for any other cultural or social activities which may involve the town as well as the University.

The Collegio dell'Aquilone includes the Serpentines, where eight single rooms form a set with common room, kitchen, bathroom facilities and a hanging garden. In the other sections, there are double bedrooms, including a common room area, kitchen and bathroom.

The collective amenities are all in the central part of the Collegio and comprise various types of common rooms, a shopping centre, a large theatre area, a library with reading rooms, book store, print centre, seminar rooms, facilities for restoring and binding books.

This library, which will become the second central library in the University, along with the large premises of the Collegio del Tridente, the theatre spaces and the seminar rooms located in different places, all form a 'bridgehead' on the Colle dei Cappuccini for cultural and social activities which the University carries on in the town centre, thus establishing a continuous interrelation between the two poles of the University.

It is envisaged that the 'bridgehead' will be particularly busy during the special courses which are held mainly in late summer and early autumn, when the regular students are on holiday. But it is hoped that it will also play an important part during the academic year and become the scene of either very specialized or very general cultural activities, in both cases attracting townspeople as well as members of the University.

8 OBSERVATIONS ON THE PROJECT

My remarks about the 'bridgehead' have led me out of the objective sphere of information into the subjective sphere of the intentions and expectations underlying my design.

I would like to continue along this track with some further comments, which will serve to round off this piece.

My first observation concerns the structure of the web of buildings. The three new residences, like the Collegio del Colle, are not separate and distinct buildings, with set boundaries. Rather they are 'territories' within a built-up continuum which comprises them all. Each 'territory' is different from the others in its organization of space, its articulation of architectural forms, the presence of some particular activity which gives it a specific role. Yet all the 'territories' have an underlying similarity that brings them together and merges them with one another.

The same thing can be observed in Urbino's ancient centre, where the quarters which take their names from the various city gates are made up of different parts, each of which is also given a name associated with some local feature. Each part is immediately identifiable by its own characteristics and yet one also feels that it belongs to a certain quarter together with the other parts.

My second observation concerns the relations established between the web of both the man-made and the natural fabrics. The introduction of such an extensive set of structures into the area of the Colle dei Cappuccini meant widespread loss to nature, which here formed an extremely valuable pattern.

So it was necessary not only to make good this loss to nature but also to try to make the new pattern as rich in quality as the one which it replaced. This aim was deeply influential in determining the relationships between the form of the volumes and the form of the land, particularly in the way that the slopes were followed or contradicted, the almost constant arrangement of gardens on the roofing of buildings, and the design of a great range of sources of natural light falling from above so that the walls would interfere as little as possible with the continuity of the landscape.

My third observation concerns the building materials. The principle of the unity of materials, found throughout Urbino's ancient centre, has also been followed in the construction of the residences, with every part built of brick and reinforced concrete. Concrete, like stone in earlier periods, has been used only where it is indispensable, either for structural reasons or else to emphasize especially significant volumetric articulations. Everything else is brick.

Concrete, of course, is an amorphous material but it is possible to give it a semblance of life by working on the moulds in which it is cast. The boards for the moulds were chosen carefully to make sure they were rough and would leave the imprint of the human labour involved in dressing them. As for the bricks, the hand-made ones used in Urbino at various periods in the past were no longer available. Industrially manufactured ones had to be used, and these lack variations in colour and irregularity of form, but care was paid to their bonding and the quality of the jointing so as to enhance the way the light glanced off their surfaces, which otherwise would have been lifeless.

The last observation I want to make concerns the image of the Centro Universitario. A purely academic architectural image is already wholly defined as soon as it is drawn. It can be published immediately because all there was to say has already been said in the drawing. I think that those architects who devote themselves to this kind of exercise are quite right in saying that construction and use can only worsen their designs. The opposite occurs with non-academic images. Their completion takes many years because it has to be nurtured by the transformations produced by use and the processes of integration set up between the artefact and nature, between building and environment. The photographs illustrating Urbino's student residences give a partial idea of what it will be like in a few years time, when their potential will have begun to develop and their image will begin to merge with all the other images surrounding them.

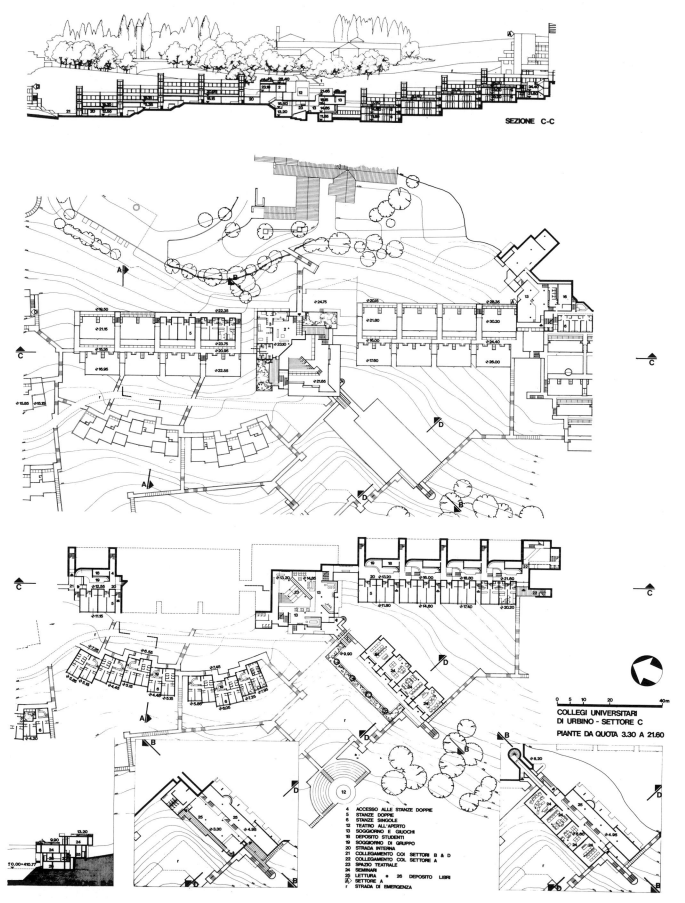

SEZIONE C-C

COLLEGI UNIVERSITARI
DI URBINO - SETTORE C
PIANTE DA QUOTA 3.30 A 21.60

4   ACCESSO ALLE STANZE DOPPIE
5   STANZE DOPPIE
6   STANZE SINGOLE
12 TEATRO ALL'APERTO
13 SOGGIORNO E GIUOCHI
18 DEPOSITO STUDENTI
19 SOGGIORNO DI GRUPPO
20 STRADA INTERNA
21 COLLEGAMENTO COI SETTORI B & D
22 COLLEGAMENTO COL SETTORE A
23 SPAZIO TEATRALE
24 SEMINARI
25 LETTURA • 26 DEPOSITO LIBRI
Ⓐ SETTORE A
r   STRADA DI EMERGENZA

Collegio dell'Aquilone
Left *section and plans*
Right *the Serpentines seen from the
Collegio del Tridente*

Collegio dell'Aquilone
Top *view showing the Convento dei Cappuccini in the background*
Left *open air theatre*
Above *internal gallery*

Opposite page *descent to the Collegio*

Collegio della Vela
Above *internal gallery*
Right *internal view of the theatre (still unfurnished)*

Ralph ERSKINE

DEMOCRATIC ARCHITECTURE: THE UNIVERSAL AND USEFUL ART

MY SEARCH FOR AN AESTHETIC

A PERSONAL HISTORY

Let it be understood that I was born to the life of an architect in the intoxicating days of the Revolution, of the battle for a new and better world of architecture.

It was an inspiring battle against the out-dated and suffocating dogmas of the academic architecture. Irrelevant for the needs of a changing world were the battles of the styles — for new-Renaissance or neo-Gothic, for the organic forms of Art Nouveau or the nostalgic romanticism of Arts and Crafts — of Lutyens, of Baillie Scott and Beresford — however artistic they might be.

Clearly these were experiences which left their mark, and it is perhaps therefore that it is with limited enthusiasm that I consider the changing styles and fashions of the Post-Modern movements. *Plus ça change plus c'est la même chose* says a French philosopher or cynic. A *bon mot* indeed but not necessarily an axiom. Our belief was for relevant and continuing change and we sought an underlying philosophy which would unite modern life and architecture and guide our attempts to acts of creation.

We fought for the new world of Modern Architecture and for functionalism as we understood it. It was at that time an architecture of analysis, but also with strong emotional ties to the new techniques and materials of the industrial age, and to Cubism in the Arts. We fought for freedom, and our faith was strong, but in retrospect it would seem that our understanding was naive, our analysis limited, and that our freedom from the old styles was rapidly and willingly exchanged for the dictates of the new.

In Sweden I found a wider belief more in accord with the early functionalism of Central Europe, a faith which included the search for a new and juster and more humane society as well as for satisfaction of the intimate daily needs of families and old people and small children.

I also found a less dogmatic attitude to style. Like Bartok who drew inspiration from the folk music of Hungary when he created his music of the future, the new and fresh and inspiring architecture and industrial design of

the golden age of 'Swedish Modern' was — without taint of nostalgia — rich with the experience of a long history of form. A history of the useful and beautiful use of the simple materials of a poor country without access to the luxurious gold and brocade and spices of the Orient. Beauty had been created by peasants and artists with wood and glass and iron, with flax and wool, and with colours extracted from the herbs and leaves of their pastures and forests. The best of Swedish Modern was an optimistic culture of the future which, without false drama, lived in the context of that tradition, in a subtle and inventive relationship with the continuity of time and place.

In much it was also an architecture of democracy, blond, open, light and accessible. It was far from the splendid and monumental symbolics of power and wealth from the history of Europe or Egypt, symbols whose meaning and potency were correctly discerned and universally used by the mighty banks and corporations of my youth, and by the latter-day dictators of Moscow, Rome and Berlin. Not for Scandinavians was the Germany of Wagner and Schinkel and Speer, but that of Bruno Taut and of the Bauhaus, of Käthe Kollwitz and of Brecht.

Here in Sweden and Scandinavia I found moving and relevant principles of design and aesthetics which are still with me.

Later history has not changed my belief. Swedish economy expanded and the drive to modernize industry, the movement to the cities and the ambition to eliminate all slums since good dwellings were considered a basic right, led to an enormous need for buildings. Quantity was achieved, and this country of eight million people built a million dwellings in ten years. It was this generation of architects and others who with their social pathos laid the foundations of a state housing policy which had never before existed, and which contributed to banishing housing misery, which was amongst the worst in Europe.

But techniques changed, 'conventional wisdom' and short-sighted perspectives prevailed, and the 'Henry Ford Method' was used. The early functionalism's interest and simplistic planning for the techniques and materials of the industrial age were seized upon, but the so-essential subtleties of the philosophy found no favour with the building industry or clients. Architects battled or — willingly or unwillingly — accepted their alloted role! Thus developed that which involved architects in despair named 'Production-

Orientated Architecture', or 'International Style' – styles which survived until they, in recent years, were put in doubt by public outcry and rejection.

THE ROLE OF AN ARCHITECT

The role of a creative architect is not in the practice of Styles, but poetically and truly to satisfy human needs, to do this with honesty in form and technique.

In my own work I have sometimes been accused of organic architecture, an accusation which, for me, is a matter of some, although not exaggerated, concern. Should organic tendencies be discerned in my design they are not to be looked upon as important or as a proof of success or failure, but as an aesthetic choice and a tendency of my personality. Other architects could, with success, have found other expressions for the needs with which I have worked.

But it would lead me yet again to realize that the precise and concrete art of architecture is paired with uncertainty on the part of architects in interpreting the language of its aesthetics, and that I must endeavour, with written and spoken language, to clarify the objectives of myself and of the group with which I work, and give my own evaluation of the success or failure of our efforts.

THE CONCEPT OF FUNCTIONALISM

Since I would consider myself a latter-day functionalist, I must define my understanding of the concept.

It must be clearly understood that functionalism for me is no style but a method of thought, a work-process which can increase our understanding of the activity in which we are involved. By no means should it be identified with the limitations of understanding or with the plans and the styles of its earlier years.

Then were the early attempts at systematic application to planning and architecture of those thought-processes which have ever expanded our understanding of the physical world, and the ideals included social, aesthetic, housing, political and scientific elements. The limitations of the Charter of Athens and other concepts, as well as the impossibility, at that time, of foreseeing the deep-going changes of the following epoch and their consequences are but one aspect – although typical – of the diverse processes which have led to the great problems which beset our age.

Functionalism, as I understand it, may not be discarded in favour of the uncertainties of mysticism or dogma. It must ever be widened and deepened. Hypothesis and invention, experiment and careful checking of results must follow one another, and knowledge be sought in all the disciplines.

AESTHETICS AND THE CREATION OF ARCHITECTURE

Knowledge of the principles of the experience of aesthetics in architecture must be sought in history, in our own experiences and in the humanistic sciences, and each method is a necessary complement to the others. The value and meaning of each must be well understood. For example the lessons of history.

Memories indeed are part of our heritage, but memories must be complete. We see the beauty of the posts and beams of Greece, of the arches of the Arabs and of Europe. Consciously or not, we realize that they spring from limitations and inventions in the art of building, we feel the completeness and the dignity of their role. Deprived of their vital purpose of most aptly and economically supporting the imposing weight of a building they become partial memories. Impoverished they falsify the reality of architecture, that most concrete of the arts, and imitate the so-different purposes, the revealing magic and pregnant illusions of sculpture or painting or theatre – of those arts which deepen our understanding of realities by evoking that which they are not.

In architecture such may become 'style', or be enjoyed as the gaiety of occasional 'follies', as an exceptional spice but never the daily bread of our cultural diet.

In music the careful study of sound, of harmony and of rhythm has expanded, not limited, the scope for intuition.

We must, as the musicians, attain greater understanding of the experiences of our art. Of the principles of composition, of the different realms of harmony and contrast and when and why they should be used. Of the formation and experience of indoor and outdoor 'rooms', the impact of light and shade, of form and materials, of texture and tactility, of our experience of structure, of real or subjective lightness and weight, and of rhythms and how they lead to stimulation or tranquillity in the human soul. We must learn of sound and acoustics and how they change a room, of how the presence of people will transform the abstraction of 'space' and of it create the warm presence of 'place', and as in a 'ballet of life' we must understand their movement through the spaces we create.

We must realize the impact of our heritage of past history and how the continuity of time can be acknowledged without impairing our duty to project the history of our own time and of the future. We must cultivate a mature judgement of where to be modest and subservient to an existing landscape or urban environment and when to enrich them with new accents. We must be able to understand the differing roles of romantic complexity and warmth such as is to be found in the cities and buildings of the Middle Ages as well as of the so-different disciplines of 'formal' composition. We must forever expand these understandings into new spheres and train our sensibility in their use. But above all we must learn all this as the essential skills of our trade: open-ended and without predestined links to 'style' – to be used for the spiritual enrichment and fulfilment of our essential role – as servants of the needs of humanity.

THE PRACTICE OF ARCHITECTURE

A COMPLEX OF CONCEPTS

Many are the facets of architecture, and we must with sympathy find understanding of a multitude of most different concepts and needs. We must, with others, enjoy a complex process of evaluation and develop a valid system of priorities.

We must achieve a delicate balance between perfection and compromise, between the so-different and often conflicting demands of practical function, of technique, of spiritual and social needs, of finance and economy (so often in conflict!) and of administrative or political organization.

None may be neglected, all must be resolved – and the result must always be checked against an overriding philosophy of the purpose of man on earth if the highest potential of architecture is to be attained.

The ambition, understanding and competence of architects varies as do those of the clients and others with whom they are involved, and tradition is no longer a certain guide as in the past ages of slow progressive change, a limited choice of materials and few and well-proven techniques.

Great is the temptation to simplify and allow a selected priority to dominate our work, be it function, technique or 'economy', aesthetics or political opportunism, with the consequence that we so often see incomplete and unsatisfying architecture, and so seldom feel that the whole task has been happily resolved.

ARCHITECTURE IS THE ART OF COMMUNITIES

FRAGMENTED COMMUNITIES

To plan a good community is a demanding task! With understanding and intelligence, maturity and enthusiasm it is difficult enough. Without these it becomes almost impossible.

To build a house, a school or a factory is insufficient, each building is as a brick in that complex edifice 'a town'. Analysis of a community leads to the realization that it consists of dwellings and shops, of places for work, education, meeting and recreation, of systems for communication and for a multitude of other human needs. Such analysis has often been made, but naively or in the interest of 'rationalization' the analysis has been built without transforming it into a meaningful whole. Each function has commonly received its special solution but the very rationalization of the particular task has often isolated it and given it a limited value for its most vital role – of contributing to the building of a good community.

Dwelling areas have become places where one merely lives and the transport apparatus can move us and our goods quickly and sometimes in comfort, but with so many negative effects on the surrounding environment that it must be isolated as far as possible from the rest of the community.

Commerce has progressively rationalized the handling of goods but has lost its vital role of giving rise to stimulating human contacts and ceremonies. It has become isolated in large anonymous shopping centres where the loss of social communication within the premises has become aggravated as the sterile motorized deserts of our physical communication have built barriers between the life of our homes and families and the joys of the market place.

Schools and universities have become bigger and bigger ghettos for a single age group, and industrial rationalization has largely concentrated itself on the production process and lost the experience of meaningful work and that direct contact between wife, man, children and neighbours which existed when human beings were active in older communities.

It has been suggested that in many industries the transport of items of production to people in their local environment could be as, or more, rational than the present transport of large numbers of people to factories.

A HOUSE IS A COMMUNITY – A COMMUNITY A HOUSE
AN ALTERNATIVE

Were it so that the stimulating human contacts of pre-industrial communities were one of the important aims of processes of production and work, then the above technique would be of great interest, buildings and plans would change, and a new, stimulating and surprising aesthetic would arise.

What consequence would such human-orientated objectives have for housing, for the construction of communication lines, for schools and universities, for trade and recreation and for the size of our townships? And for the character and our evaluation of architecture!

It would seem to me that in the intimate interplay and confrontation of different insights, interests, generations, subcultures, activities and situations of varying size there are essential values which are commonly lost in our modern communities. In this, rather than in contrived and romanticized manipulation of form, would arise the complexity we seek and new forms would appear.

Instead of housing or working areas, it would be important to create 'places for living' which offer varied life styles, parts of towns where dwelling, work, study and recreation take place in as close contact with one another as possible. This is equally important whether we be involved in an individual building or a community plan.

The Goodman brothers have conceived of a community where meaningful work and varied experiences are amongst the main objectives for the culture and planning, and have established that such communities could not be too large. They suggest that our modern cities have grown with very different priorities in mind.

New communities and parts of communities must be built for the rapidly increasing world population. How important it could therefore be if it were possible, together with intelligent, knowledgeable and interested people, to plan complete and proper 'places for living'. Useful, compassionate and poetic places. And how similar in concept, although not in form, they might be to the older villages and towns we know.

It is my experience that much can be learnt from research, discussion and literature, but equally much from experiencing, observing and analysing the interplay between built form and present-day life in such older townships and villages. Likewise that whilst it is this interplay which is so relevant, this realization can readily be confused by observation of the beauty of their style and the detail. For these are often the expression of the economics, techniques and beliefs of another culture and age, and imitation will lead to the falsifications of nostalgic pastiche.

Furthermore I must admit, that whilst I have been fortunate enough to design small communities or parts of communities, it has despite all my endeavours not yet been possible to achieve the weave of functions of which I speak. These projects are largely 'dead' during week-days and week-end places at week-ends and the so-essential richness of life has not arisen. The charm they may have could be the 'Aesthetic Trap' artistic manipulation which gives intimacy and personal situations but environment which, although improved, is still thin in content. Not community-places but the monofunctional housing areas I decry and a palliative rather than a solution for our living environment?

FOUR UNACKNOWLEDGED AND POWERFUL GENERATORS OF CHANGE

With some justification it could be said that it is not architectural philosophies, which are today the important instruments of change, which affect my architecture but the insights of scientists, economists, philosophers, authors and many other opinion-formers interacting with national and international institutions of political and economic power. The special dynamic for architectural change has come when such insights have been formalized in building — or other laws.

FOUR FACTORS
For the present purpose I will select four factors as typical important agents of this architectural change in Sweden.

1 Democratic participation in decision-making processes
The 'user-client' brings new insights and evaluations which can fundamentally differ from those of the traditional 'sponsor-client', and architecture and the architect must change. New qualities must be discerned, given form and defended.

2 Minority rights (at the moment especially of the handicapped)
The aesthetic stimulation which arose with the manipulation of varied floor levels and stairways becomes impossible, and there arises a different aesthetic of extensive horizontal floors, with ramps and lifts at any unavoidable changes of level. Serious consideration of the needs of children, the aged, immigrants and other subcultures and minorities would introduce further new form-elements in architecture and planning.

3 Economy in the use of energy
It is realized that energy is a valuable, scarce and polluting resource. Building volumes must therefore become simple and heavily insulated in both hot and cold climates, thermal bridges must be minimized and windows severely restricted in size. But controlled solar-heat collection can lead to carefully orientated glass areas of considerable size.

The architecture of glass buildings however beautiful, is recognized as symbolic of a naive, wasteful and irresponsible culture, and laws on energy conservation have made such architecture impossible. A new, wise, responsible and beautiful architecture must be invented.

4 Economy in the use of all resources
The first three factors have to a considerable degree been ignored by architects in the past. The fourth, always operative for most utility buildings, can also be considered as a fundamental condition of human rights in a world where the limited resources are so inequitably distributed between classes, races or nations. With the spread of media-communication the tragic effects of such inequalities may, as well as being an intolerable injustice, become a threat not only to the underprivileged, but also to those of us who waste resources in the wealthy and privileged parts of the world. For those privileges can be violently challenged by the deprived majority of humanity. A subtle and inspired architecture and aesthetic of economy should therefore become the overriding interest of architects in their professional role and the wise use of the saved resources their concern as citizens.

COMMUNITY PROJECTS

JÄDRAÅS, 1951

A forest village group originally for lumbermen and railway workers, now used by commuters.

State-loan 'low-cost' housing with plan and aesthetic characteristics intended to fit the special needs of the users and the forest situation.

Scheduled as protected environment.

HAMMARBY

Small industrial community for 1,000 inhabitants and pulp industry.

Community plan, state-loan houses, infil, flats, homes for old people, rehabilitation of historic buildings, school extension, community functions, shops, water tower, industrial buildings, etc.

Full participation of users on all work done during thirty-five years. Both old and new buildings scheduled as historic monuments.

STUDLANDS PARK, NEWMARKET, 1969

Housing for sale. Very low cost achieved by very simple volumes, construction and materials.

Full traffic separation, my first introduction of garage courts with arrival points. Main centre, small group centres, light industry.

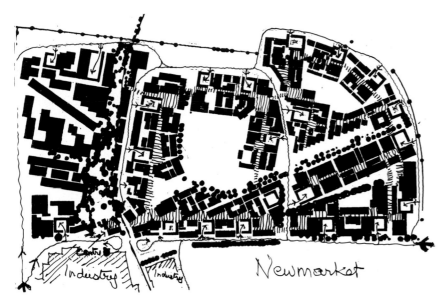

Newmarket

KILLINGWORTH

Housing at edge of lake.
Only part of scheme built according to our design as shown in the model.

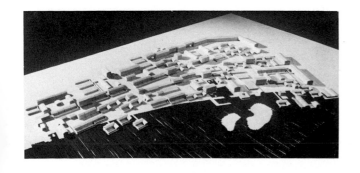

BRITTGÅRD, TIBRO

State-loan mixed development, 350 dwellings with varied ownership – rental, residents' association and private ownership. Flats and houses.
Windscreen buildings, simple volumes.
Separated prefabricated balconies and other elements.

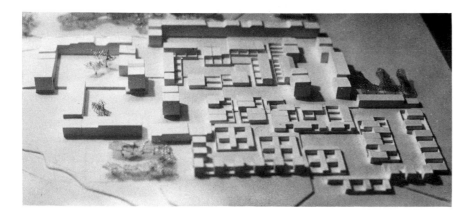

BARBERAREN, SANDVIKEN, 1962

Mixed development, flats, offices, health centre and shops in city centre.
Small city character and subarctic elements in design.

Mixed housing – old people and families

Kindergarten

Youth centre

BRUKET, SANDVIKEN
State-loan development for 700 dwellings and other functions.

BYKER, NEWCASTLE-UPON-TYNE

Flats, houses, homes for old people, shops, schools, doctor's practice, community functions, etc. in extensive landscaping with smaller and larger play areas, and street furniture of different types.

In different areas such as the top of the hill, slopes, flat parts, etc. social groups have been formed and given individual character and local characteristics within the general grammar of the project. In this way uniformity has been avoided.

Continuous buildings protect against traffic noise – and northerly winds. Within the scheme we have planned for traffic separation, for social reasons and safety.

The whole of our project is state-loan development, the final stages have, with change of government policy, been sold to private enterprise, and it seems doubtful whether the environmental qualities will be maintained.

Vernon Gracie, David Hill and Ralph Erskine work with local children

81

THE IMPACT OF CLIMATE

When considering the problems of building in the north, to talk of an architecture of climate would be to tell only half the story. It is people in the climate, the cities and the landscape, people alone or in families or crowds that count. Ordinary people, not architects, people who sometimes are born in the north and know it and love it or hate it, other people who are moving from more popular areas to small isolated communities in the wilderness, and who must be given the amenities they previously enjoyed.

The people of the chilly winter greyness and verdant summers in the more populated areas to the south could find insights from studies of the extreme conditions of their northern neighbours.

Man in his ingenuity has invented many ways of protecting his puny body — of maintaining its surface within the narrow range of temperatures and humidity which allows for survival. As his inventiveness and artistry increased he has moreover created conditions of convenience, comfort and pleasure. With time he has in the south created cities and buildings which are works of art and a witness to the genius of human cultures.

It can be difficult enough to express thoughts in words — how much more difficult to say them in concrete and wood, in asphalt and grass, to say them with precision and warmth but without unnecessary pathos or exaggeration. I hope that we architects can help to give dwelling a form, to make a place with a potential for contentment. But in the final count it is the inhabitants who will give the same dwelling its meaning and will change our architectural *space* to *place*!

VÄXJO FLATS, 1954

An early attempt at subarctic technique and aesthetics in middle Sweden, with prefabricated concrete construction.

Simple enclosing volumes, restricted window sizes. The balconies hang from the roof to avoid the cold bridges that occur when a concrete floor is continued outside the building envelope.

VILLA AT LISÖN

A subarctic house on Baltic island near Stockholm.

The half sphere gives maximum heated volume with minimum surface area and heat loss.

It is sunk into the ground to seek further protection and is surrounded by sunny wind-protected terraces for spring and autumn pleasure.

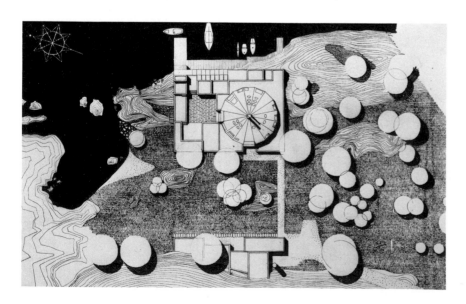

VILLA LIDINGÖ, GADELIUS

Subarctic architecture in Stockholm. Protective house sunk into the ground to the north, open to the south, roof garden, separated summer structures.

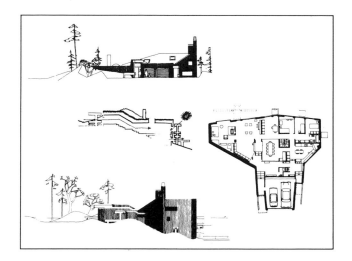

VILLA STRÖM, STOCKSUND

Subarctic in Stockholm. Heated cube, with maximum volume, minimum surface; boiler in middle with winter-garden above, reflectors on roof for low winter sun – separated balconies for summer use.

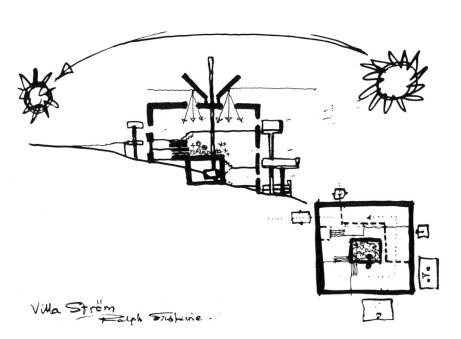

OWN HOUSE AND OFFICE, DROTTNINGHOLM, SWEDEN

Stockholm subarctic architecture, with simple technically and aesthetically enclosed volumes, small windows, free roof to avoid ice formation; separated outdoor structures, wooden bridges at entrances ease snow clearance. Red granite garden reduces maintenance and accumulates sun heat in sunny wind-protected outdoor spaces. Tall chimneys improve function and emphasize importance of heating in cold climate.

Free tropical *interior*

BORGAFJÄLL HOTEL

Exceptionally low-cost sport hotel – largely as a result of designing the building, its details, lighting fittings, furnishings, etc. so that they could be made by local labour with simple and very unpretentious local materials and techniques.

Building sunk into ground and snow for protection, ski slopé on roof, sunny wind-protected balconies. Free forms and volumes in the *tropical* interior.

PLAN FOR CENTRE OF KIRUNA AND CONSTRUCTION OF CITY BLOCK

I was first commissioned to carry out a planning study for the city centre. This I gave northern characteristics. A conventional plan was executed – one block was built according to my design.

Subarctic architecture in the subarctic with wind protection, sun-catching environments, alternative indoor–outdoor communications, well-enclosed and protective volumes with rather small windows, indoor balconies, small outdoor balconies for deep-freezing of fish, venison, etc. State-loan housing.

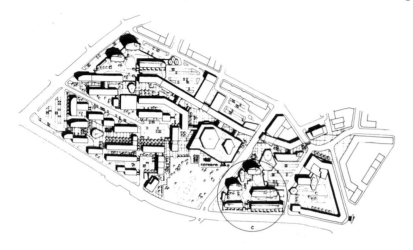

LULEÅ SHOPPING CENTRE, 1956

My speculations about summer and winter life in the north and the survival potential of sociable street-life during the long, cold and dark winter led me to plan indoor public spaces – and the world's first indoor shopping centre was therefore built.

The building's interior has largely been replanned by a new owner (an architect) a few years ago.

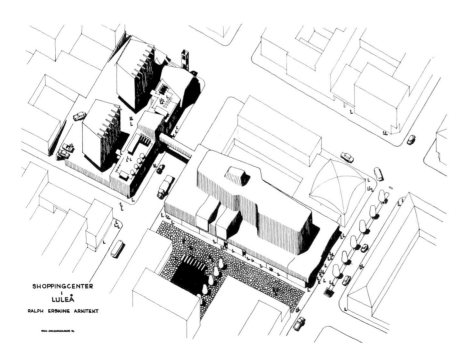

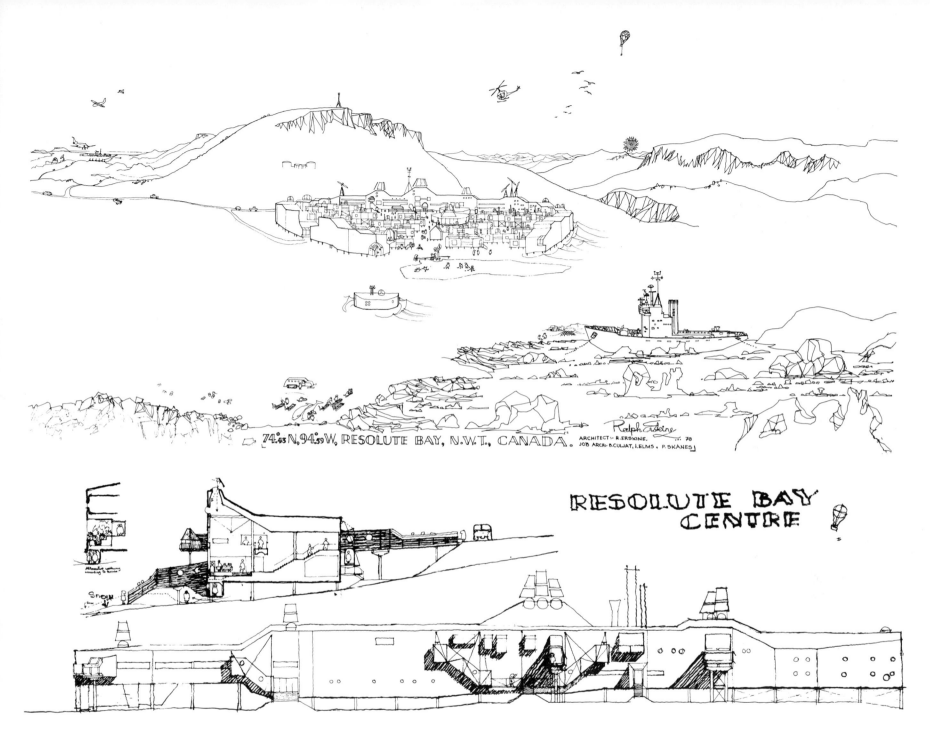

74.°43 N, 94.°53 W, RESOLUTE BAY, N.W.T., CANADA. ARCHITECT:- R.ERSKINE. 70
JOB ARCH:- B.CULJAT, I.ELMS, P.SKANES

RESOLUTE BAY CENTRE

RESOLUTE BAY, ARCTIC, CANADA

An arctic community for 700 Eskimos and south Canadians incorporating many concepts of social planning and climatic design — only partially constructed. Southerners, and especially Eskimos, were involved in the whole of the planning process from the choice of site and form of the township to the placing, form and details of the buildings.

COLD AND HOT CLIMATES

Intensive impressions during visits to Arab countries and southern states in the USA have combined with the results of my attempts to analyse the consequences of cold climates for people's lives, towns and buildings and I have become conscious of major parallels between the hot regions and the cold.

Since cold and heat are experienced so differently and few people have lived both in arctic and tropical regions, it has not been observed that for their effect/potential on architecture and planning the similarities between the cold arctic and the hot-dry desert and savanna region are extraordinary. With the hot-humid climates the similarities are also important, but are not always equally obvious.

IN CONTRAST THERE IS SIMILARITY.

In the arctic it is important to catch the sun and avoid the breeze, in the heat it is equally important to avoid the sun and catch the breeze. The arctic is a cold white desert with drifting snow, in the tropics there are hot yellow deserts with drifting sand. A warm place in the subarctic is an oasis with trees, a wet place in the desert gives the same result. Except where there are special resources, communities in both are usually small and isolated and the traditional cultures have been nomad. Survival techniques and the whole culture have in both been highly specialized and directly related to the impact of the extreme climate. Today architecture and planning that are properly considered would show certain striking differences but there would be very much that is strikingly common to both regions.

Simple volumes of well-insulated buildings will arise in both hot or cold regions. They will have restricted window sizes and be surrounded by light structures where the pleasures of the more temperate seasons may be enjoyed.

TOWNSHIP IN A COLD CLIMATE (NORTH OF POLAR CIRCLE)
South-sloping site, town plan and the form of buildings help to catch sun heat and protect against cold northerly winds.

Housing block in Svappavara. *The bold roof screen protects the balconies from northern wind and increases the sun-catching area.*

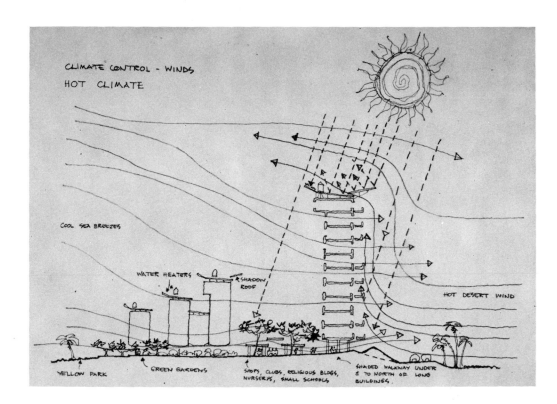

CENTRAL PART OF TOWN IN A HOT CLIMATE
Orientation of buildings for maximum shadow hours (roof projection important). Sea breezes from west penetrate plan. Desert breezes from south-east hindered.

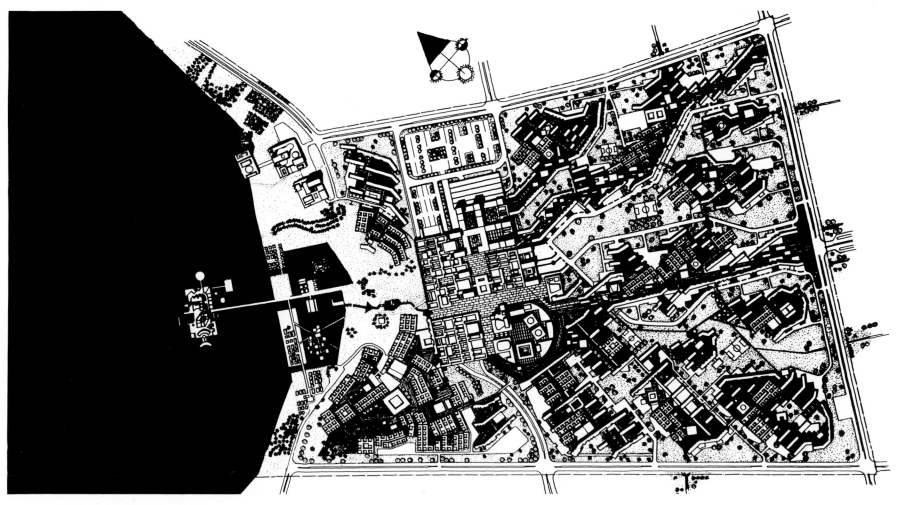

DIVERSE PROJECTS

GYTTORP COMMUNITY AND SCHOOL

Community plan with housing, shops, centre and school for a small industrial community. 'Low-cost' state-loan development.

A small-scale, intimate, low and middle stage school with a plan and environment which attempt to follow the emotional and intellectual development of children as through the years they move from the intimacy of home life to their life in the community at large. The first class is in a small wooden house in the forest, the last is in the school's social centre which is used by adults and children and is beside the village square.

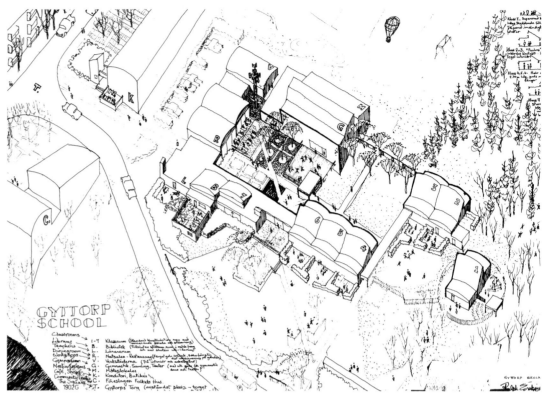

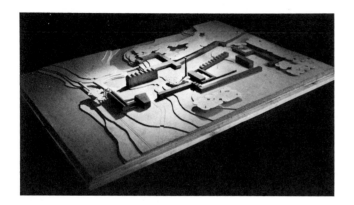

BAKERY AND OFFICES, PÅGENS, MALMÖ, 1969

In south Sweden, a less extreme climate.

Plan and furnishing give personnel the possibility of alternative work and social relationships – in groups or individual and private.

A relatively small-scale, intimate and *personnelisable* environment. A *combi office* with continual participation potential. A popular alternative to office landscape.

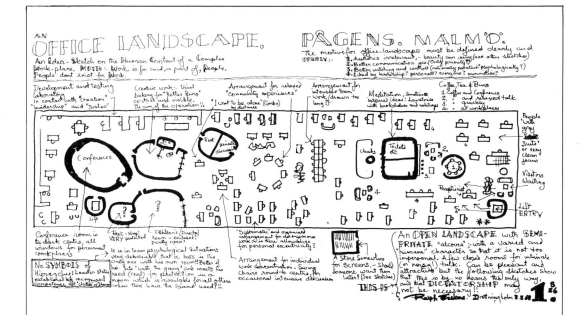

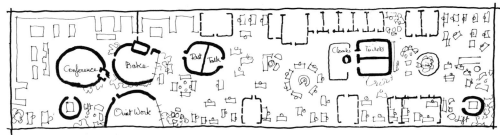

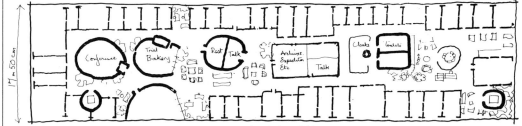

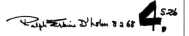

CLARE HALL, CAMBRIDGE

Post-graduate College with flats, houses, study group and social centre. An attempt at a college – architecture of good quality – useful environment without monumental overtones.

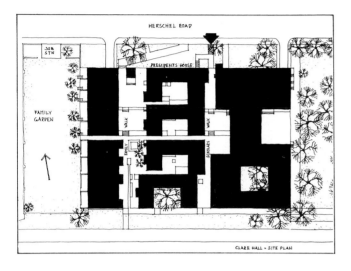

BODAFORS CHURCH

Remodelling of existing church and additions for social functions.
Simple forms and materials, beautifully executed in one of the parts of Sweden which has an excellent tradition for woodwork.

FRESCATI – STOCKHOLM UNIVERSITY
STUDENT CENTRE, LIBRARY AND SPORT HALL

Our disposition of the functions and the siting of the buildings departed from the original plan which was for a much larger university than that now envisaged.

The student centre, humanities building and library enclose a verdant university court; the sport hall is related to a hill with oak trees and to grass areas for outdoor activities. The student centre is a recreation sculpture, the library a rational building with sculptural elements, the sport hall an economical structure with pleasant social elements.

Certain common architectural characteristics have been sought between the library and the existing buildings on the one hand and the social recreation buildings on the other hand, in an attempt to create a Frescati character. These projects were designed in contact with students, professors, librarians, and those who in different roles are employees of the university.

Eldred EVANS and David SHALEV

A SENSE OF PLACE

It is difficult to define architecture, but one of the most beautiful descriptions of it comes from Le Corbusier: 'The business of architecture is to establish emotional relationships by means of raw materials.'

In our work we are concerned with Place making. We always find ourselves trying to create an entity which is both part of a larger one and a grouping of smaller ones.

We are preoccupied with continuity and progression, and that also means constant transformation in space and time. The Place in which suddenly a new building appeared is not the place it was before. The hill was always there but now you are aware of it. The old terrace was always there but now you are aware of it.

A site plan explains how the building relates to its surroundings. It explains why the building is where it is and what it is. It explains how you get there, how you get in and out, what you see from inside, where the sun is, and what the topography is.

In designing we like to think of:
Entering the building.   Entering the street.
          Entering the building.   Entering nature.
A building in the landscape.
                    The landscape in the building.
A building in the street.   A street in the building.

To us buildings are not only objects, but part of a continuum, real or imaginary, or both. By their very nature they are a microcosm of the city.

We look at the Place and try to feel its scale and rhythm, the Place as it is and the Place as it wants to be.

We try to understand the essence of the place we want to build. This has nothing to do with the usefulness of the building but only with the kind of place we want to have, its absolute being, its intrinsic nature. We are in search of an idea.

We invent the spaces which embody the idea. We manipulate structure and enclosure, size and scale, light and texture: spatial order.

We first try to find the best means for making those spaces. We then look for an appropriate technology. The skill of putting things together has to do with materials, with details, with the business of putting up a building.

A client's brief does not exist on its own as a finite document. The brief consists of a list of rooms with a few preferred relationships and some notes about the client's previous experience, and above all, a fairly clear

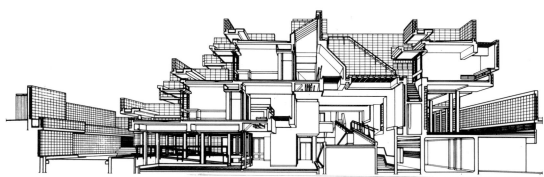

Boundary Road, master section: a cluster of dwellings in the city

idea about what he does not want. The real brief comes into being as part of the design, and constantly changes.

The design process does not start with the brief, but with an idea. The idea which represents the essence of the building.

More often than not the architecture and the sense of place are created by those spaces which were never asked for, they were not in the brief, they were never contemplated by the client. They are not whimsical, they are the essence of the building.

Courtyard	Garden
Steps	Piloti
Terrace	Roof
Street	Bridge, weir

Van Eyck said: 'What is large without being small has no more real size than what is small without being large. If there is no real size there will be no human size.'

To us the only conceivable hierarchy is that of privacy and community, of aloneness and togetherness. A public place relative to a private place is, at the same time, a private place to another public place. This is pluralistic or complex order.

The idea is that spaces are related not by their proximity, being next to each other, but by being within each other. The small 'my' space is part of the large 'our' space. The demarcation or definition of the private realm is, at the same time, the containment of the public realm. This means that Containment is not the opposite of Continuity, and Definition is not the opposite of Transparency. They are one and the same. This is also pluralistic and complex order. That is to say, that a space belongs to itself and, at the same time, to another space. Is this not the nature and the essence of our inner space?

In Boundary Road the residential quarters are within

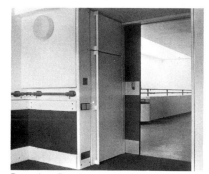

Boundary Road: an open door

Taoiseach's Residence:
a cluster of dwellings in the landscape

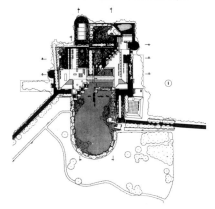

the main space. The private room is within the space of the house. (The house space is common to the private rooms, but private to the main common space – the hall.) This spatial order does not distinguish between internal and external spaces. The main space and the garden, the room and the terrace are not two related spaces but two parts of one enclosure.

To us, the elements of space are walls and light. The walls are as much a light source as the windows. Walls reflect light. The texture of the walls determines the quality of light just as much as the shape of the window. A pitch dark space is not a space. Louis Khan wrote: 'A great American poet once asked the Architect, what slice of the sun does your building have? What light enters your room? As if to say, the sun never knew how great it is until it struck the side of a building.'

So light strikes.   Light also flows.
        There is harsh light.   There is soft light.
You can touch light.
                You can almost hold it in your hand.
        Light is colour.
Light changes.
                Light can burst into space, or trickle in as though through silk.

We find that the spatial order of our buildings makes them deep, extensive, horizontal. One of the main sources of light is therefore from above, through the 'fifth elevation' – the roof. It is as if the building was built to contain light.

The choice of structure is determined by the spatial order. The choice of materials is determined by light and texture.

Our buildings want to be white because of the kind of light we want. The basic materials are white cement and white aggregate. It is the same for the form moulded on site as for the prefabricated elements. It is the same for the inside as the outside. It is the same for surfaces you walk on and those you touch or lean against.

Unity of texture and colour is consistent with unity of space. Is this the reason why we like a farm or a medieval town?

The detailing of a building is an organic part of the design process. The detail already exists in the first sketch. You have only got to discover it.
The detail grows out of the space. A space is used.
The space grows out of the detail. A detail is used.
Unity and diversity in space – unity and diversity in detail.

One aspect of detailing has to do with the skin, the edge, the transition from inside to outside.

The window is not only 4mm of glass to let the light in and keep the rain out. We can sit on it, stand on it, walk on it, work on it, play on it. It is an opening, a lamp and a radiator at the same time. It is a chair, a table and a step. A window is a Place between the inside world and the outside world.
A balustrade is a Place to sit, play, climb.
A Place to grow plants.
Steps are a Place to be, to sit, to play, to look.
Detailing is Place making.

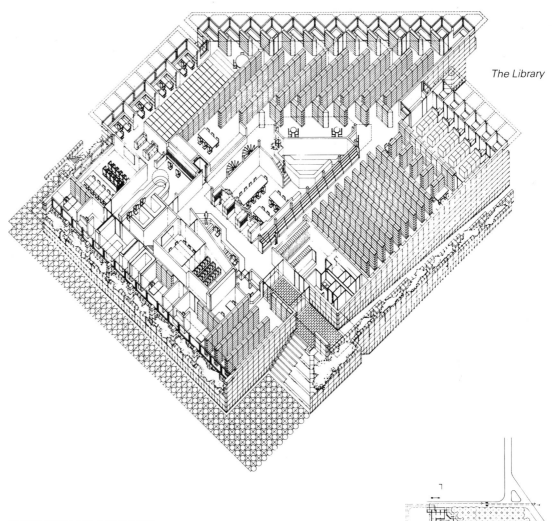

The Library

SHRIVENHAM LIBRARY
Royal Military College of Science

In 1981 we submitted this project as a competition entry to the P.S.A., for construction in 1983.

The promoters envisage a linear university plan with a pedestrian precinct and/or a formal approach avenue as a central space. The future sequence of teaching facilities to the north of the future central space may then be consistently serviced from Lower Woods Road via a series of screened car parks. The continuous zone of trees between the teaching buildings and the car park also provides a secondary east–west pedestrian route culminating in the proposed wooded area north of the library which in turn screens the car parks from the western approach.

The library, being the first in a sequence of new teaching facilities, centrally located *vis-à-vis* the Sandhurst blocks and to one end of the pedestrian precinct, seems to be in a strategic position to form together with a possible future adjacent building – the College Hall – the Circulation Concourse. The proposed Concourse thus acts as an entrance or 'gateway' to the main pedestrian precinct, when approached from the main residential accommodation to the north of Bower Brook.

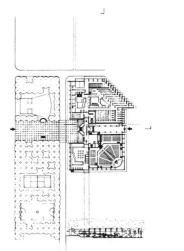

Lower and upper concourse level plans

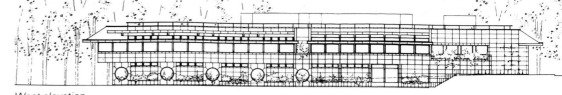

The concourse doubles up as foyer area, all as described in the brief.

This grouping of common facilities to the entire College – the student areas, the hall, the cafeteria and the library – is likely to become the hub of the College. The concourse will remain the 'main entrance' to the College but may by linked in future with a sequence of minor entrances from the north into the pedestrian precinct.

The main entrance to the library faces the pedestrian precinct. The approach to the main entrance may relate in one instance to the north–south proposed avenue of trees linking the residential accommodation with the pedestrian precinct, or otherwise to the proposed Concourse.

For day-to-day use and for easy supervision and maintenance, the library is, in the main, designed on a single level entered from the pedestrian precinct, with a lower level containing the text book library also approachable from the service/parking zone.

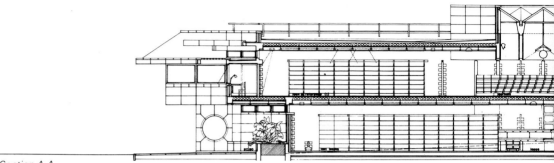

West elevation

Section A-A

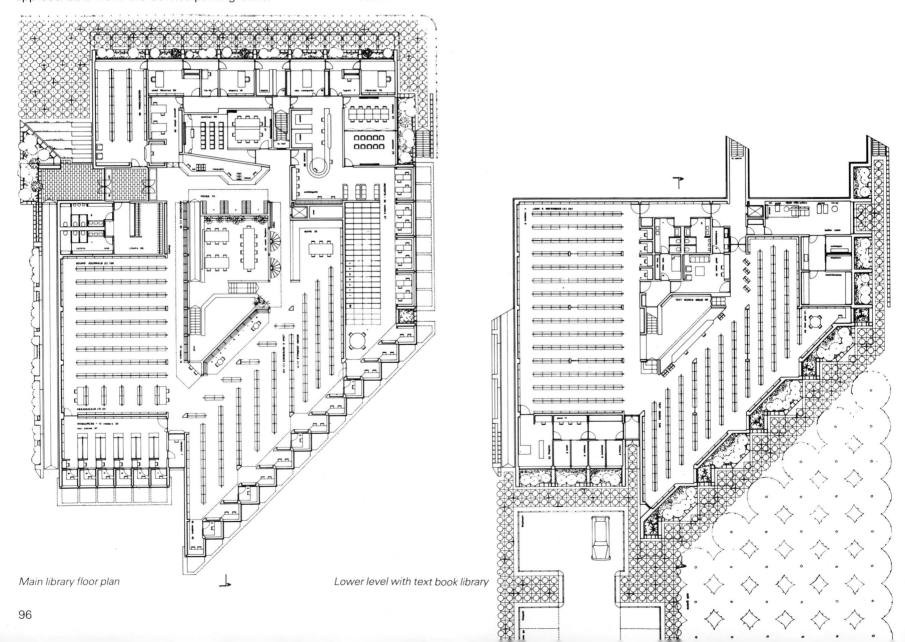

Main library floor plan

Lower level with text book library

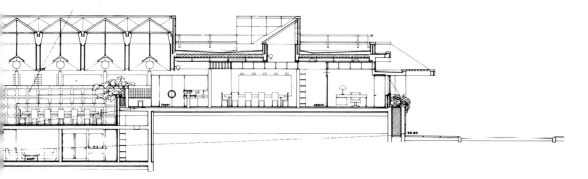

East elevation

The library is planned around a central top-lit reading room with the private and semi-private reading cubicles always on the periphery of the building with an outlook on to green areas, for peace and quiet.

The central small reading room (twenty-two readers) is formed and contained by the three main building elements. The north-west block contains the main collection stacks: main loan and reference books, loose and bound journals, the maps, the text books and separately the reports section. The east block contains all the secondary collection stacks on both levels. The south block contains the interconnected administration and control suites. Each of these three 'buildings' or zones may be modified and further developed in accordance with detailed observations and requirements.

In summary it may be said that in designing this library we tried to achieve an environment which is dignified yet not overbearing, an architecturally stimulating space, or place, with an atmosphere conducive to study.

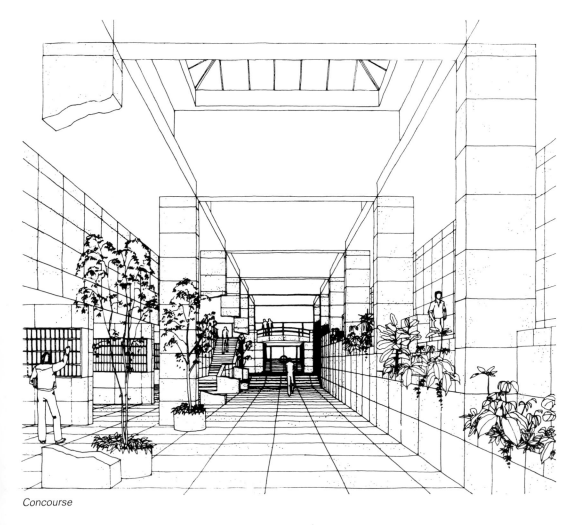

Concourse

Text book issue area

Reading room

48 BOUNDARY ROAD

The Home for the Younger Physically Handicapped which we designed in 1972, unlike many existing homes, is located in a dense urban residential area, so that the severely disabled residents can fully integrate with the local community and develop closer contact with their relatives and friends. This means that, unlike other homes, this one is built on a confined and restricted site. Consequently, many of the accepted types of single-level homes are inappropriate in this context.

The brief, compiled jointly between the client and ourselves, called for a building that should 'provide for an environment in which the residents could lead as active, independent and varied a life as their disabilities allow. The young or middle-aged handicapped person may feel that a normal life has been denied him and looks upon his admittance to a home as an opportunity to lead a fuller life ...' The very severely handicapped should be encouraged to develop interests and activities which absorb much of the day. Such facilities could also be utilized as an occupation centre for non-residents.

Whilst the planning concept acknowledges the need for intimate family grouping of the residential accommodation, it takes into account the high visual awareness of the resident who spends much of his time observing.

Apart from being a Home, a house in a garden, this building is essentially a community hall: a meeting place between a family of highly perceptive and most sensitive people who, in spite of their disability, want to lead a normal, active and creative life in the outside world.

On the other hand a severely handicapped person cannot always take part in all activities. He should however be able to participate – to see, to hear. We therefore preferred one articulated common space, bustling with life, which unfolds to the viewer in more ways than one, to the usual series of isolated functional and underused common rooms.

The Home, which will accommodate thirty residents, and the staff flats form the envelope to the common space. The two main aspects in the design of this home are the creation of a quiet and pleasant residential atmosphere, and the provision of extensive outdoor space.

The building provides for a variety of protected and easily accessible outdoor spaces in conjunction with, and as an extension to, the adjacent rooms. The ground and roof gardens are for common use and the terraces and courtyards are private. Every resident has direct access to his room or suite without having to use the main entrance.

To ensure privacy and maximum exclusion of noise, the building grows out of its garden walls which in turn contribute to the street by way of planters and seats.

All the private accommodation faces the sun, and the building presents itself to the street as a group of dwellings rather than an institution. The entire south face of the building is intended to be a cascade of plants.

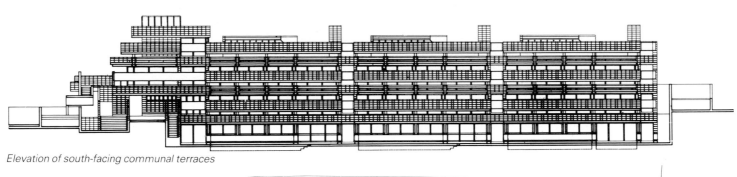

Elevation of south-facing communal terraces

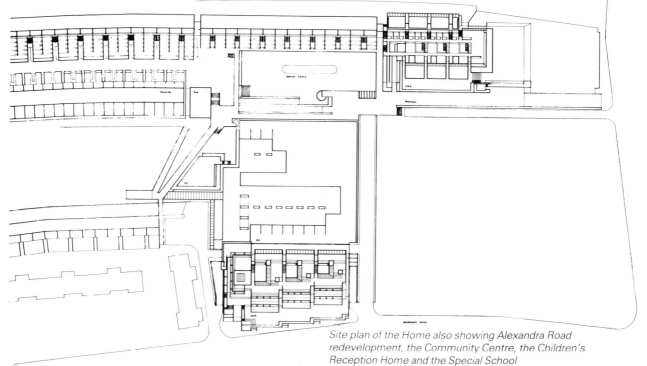

Site plan of the Home also showing Alexandra Road redevelopment, the Community Centre, the Children's Reception Home and the Special School

South-facing communal terraces

Boundary Road, the pavement

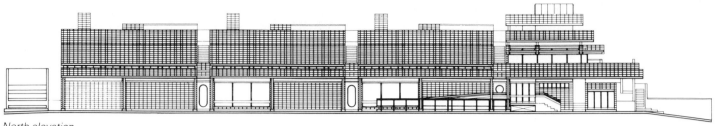

North elevation

Garden level plan

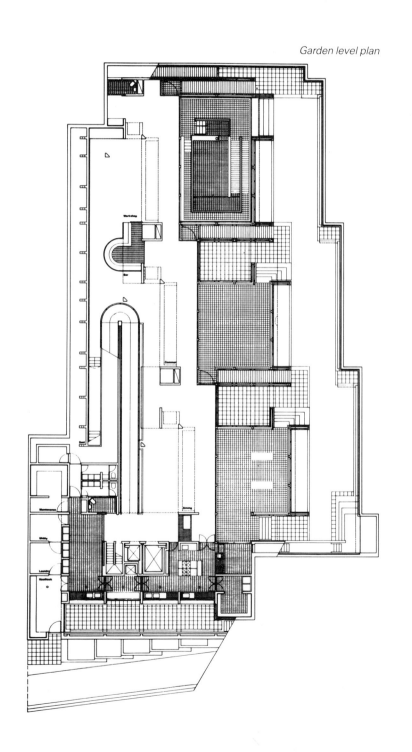

First floor plan

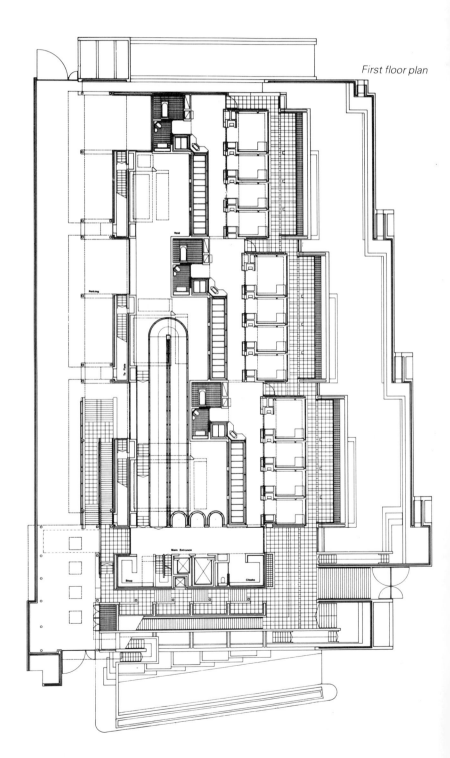

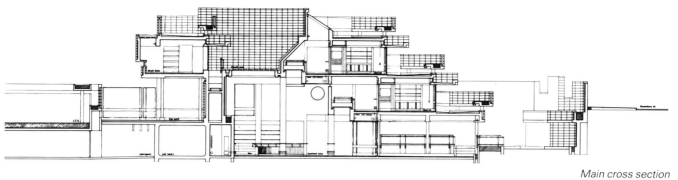

Main cross section

Second floor plan

Roof terraces plan

The house common area, shared by four or five residents and looking from the main space up into a house

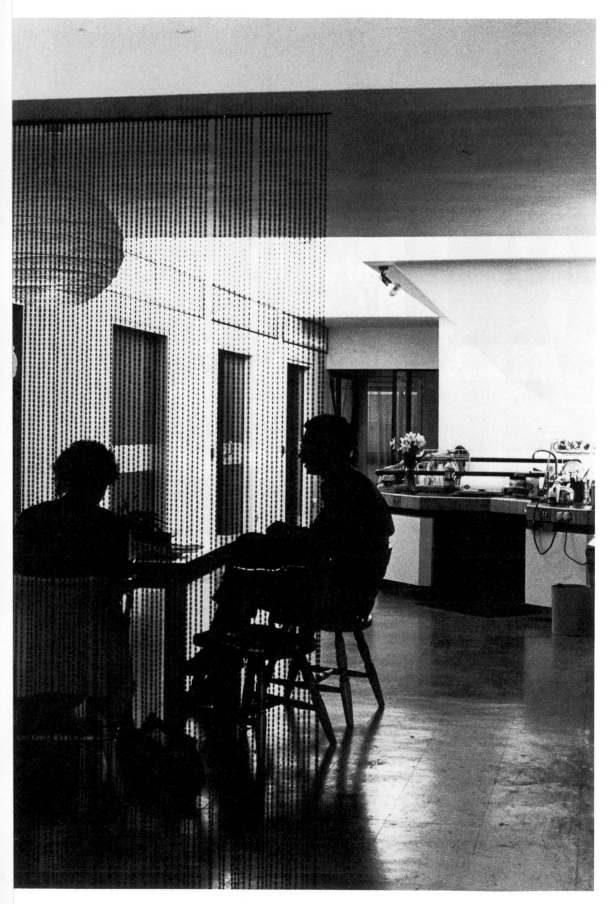

shelves and opens on to a terrace accessible from outside, and each "flat" also has a door on to the appropriate terrace.

'The rooms are not large, but living areas as well as corridors and areas throughout the building are spacious.

'Each part of the building has its own staff, with the Head of Home having overall responsibility. The busy Community Centre has its separate staff team, as does each residents' "flat", where the team is led by a senior. There are no medically qualified staff employed; some are trained social workers, but most have no previous experience of working with disabled adults. There are also a number of volunteers working, often attached to a particular "flat".

'The fundamental idea of the Unit is that residents should live independent and full lives with staff available to give the physical help needed and to provide emotional support and guidance. At the same time, residents are encouraged to take on as much responsibility for their own lives – in the widest sense – that they can manage; indeed, potential for this is one of the criteria for admission. Residents make all their own decisions; get up at their own time and in their own manner, eat what they like and when they like, go out if and when they feel like it and go to bed when they choose. The only restrictions on this freedom are consideration of other people's needs and availability of staff, and this has to be agreed, between residents and staff available at any time.'

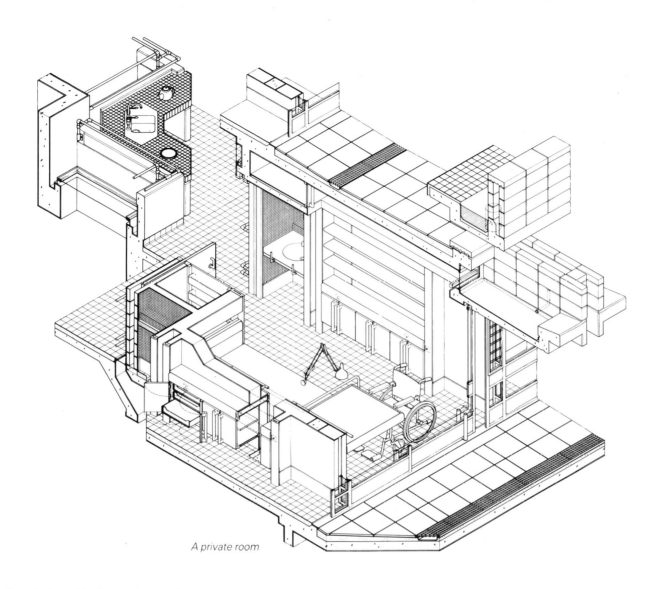

A private room

The house kitchenette, through the resident's room to the private terrace

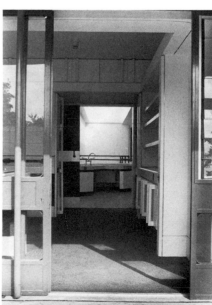

The terraces

'The link is the window'

A RESIDENCE FOR THE TAOISEACH AND A STATE GUEST HOUSE: DUBLIN

We designed this intricate complex of buildings in '79–80, as an international competition entry. The project was awarded first prize but has since been abandoned. We then wrote:
'The Taoiseach's Residence and the State Guest House, although designed as two distinct and private entities, achieve by means of their juxtaposition a SINGULAR SENSE OF PLACE. This, in our view is the prerequisite to creating a representative building of this nature.'

We conceived of the Residence and the Guest House as one integrated cluster of buildings, designed as an organic part of their surroundings, in which each part maintains its identity and privacy.

The cluster presents itself to the park as one homogeneous entity, thus achieving an appropriately dignified and prestigious appearance while maintaining within itself a pleasant and habitable scale.

An important factor in our attitude to, and interpretation of the brief was the desire to strike a balance between the FORMALITY of the occasion and the INFORMALITY of human relationships.

The uniqueness of the site, both physically and historically, is enhanced by integrating the existing features in the new cluster of buildings, namely: the North Avenue (with its walls), the stable building, the Keep and the garden wall.

The grouping of the buildings and the approaches to them established the cluster as the dominant feature in the park and creates a positive relationship between them: House and Garden: THE LINK IS THE WINDOW.

The approach to the State Guest House is via the partially realigned south road, along a new avenue of trees. The approach to the Taoiseach's Residence is through the existing (improved) North Avenue, leading to a festive arrival space formed by the Keep, the stable building and the water, off which an arbour, formed by the existing garden wall, leads to the Taoiseach's private apartment.

The cluster is formed by the loose grouping of eight distinct building elements, each of which extends into its own private and protected walled garden compatible in scale with its use. The main building elements are: the reception rooms, the Taoiseach's private apartment, the staff quarters, (all of which form the Taoiseach's Residence), the guests' common rooms, the guests' suites, the guests' rooms, the service wing, (all of which form the State Guest House) and the duty and maintenance building in the stables.

In designing the cluster we adhered to the following guidelines:
1. creating a positive and unobstructed relationship between the cluster and the park;
2. creating a sequence of useful, private and protected walled gardens, using the existing garden as the generic form;
3. giving each building its private realm while creating a coherent whole;
4. establishing a three-dimensional concept within which the formal and informal parts of the buildings will co-exist;
5. benefiting from the sun. Common areas are inward looking. Private areas are outward looking.

The proposed cluster of buildings, being located in parkland and growing out of their garden walls, adopts a spatial concept in which external spaces do not relate to internal ones by their adjacency. The indoor room and the related outdoor room are part of one and the same space.

This fluidity/continuity/transparency of space calls for glass screens (double glazed) in the appropriate places, always protected by a translucent, ventilated canopy. This consistent edge detail played against the extensive garden walling (reconstructed stone or faced masonry) gives the building its distinct appearance and profile in the park.

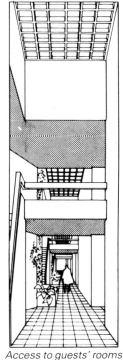

Access to guests' rooms

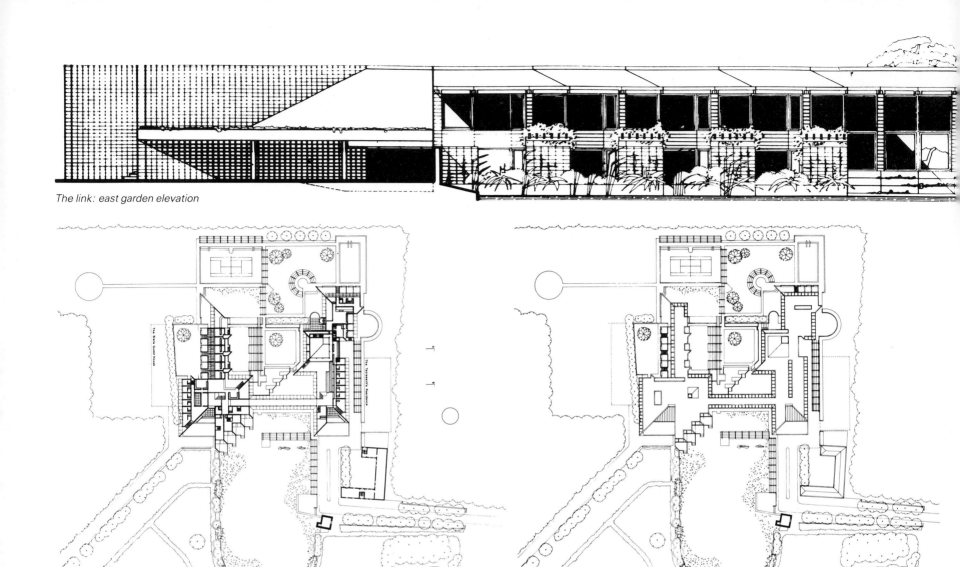

The link: east garden elevation

Plans at garden and roof levels

Private apartment

Taoiseach's Residence, reception rooms

State Guest House, main entrance

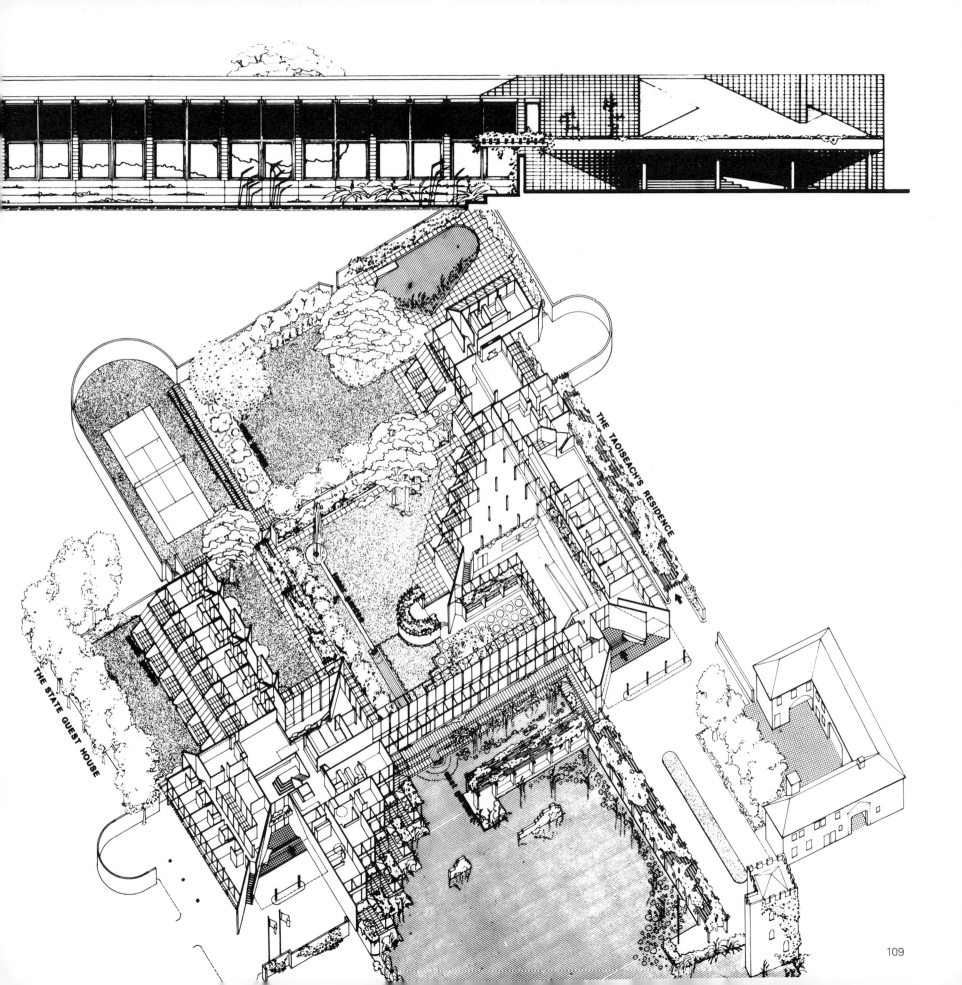

THE TAOISEACH'S RESIDENCE

THE STATE GUEST HOUSE

109

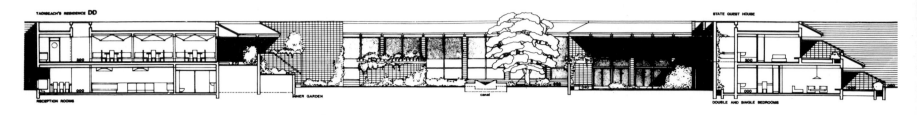

State Guest House, section and first floor plan

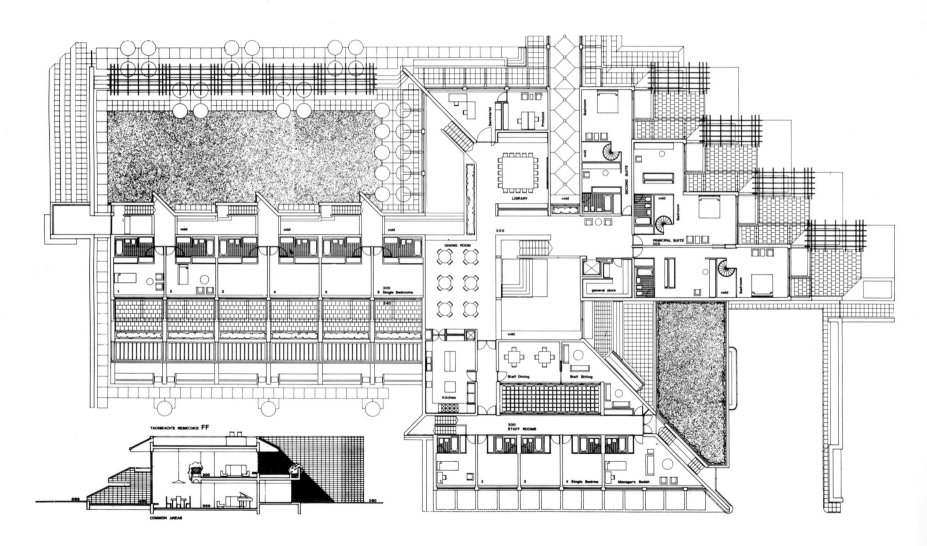

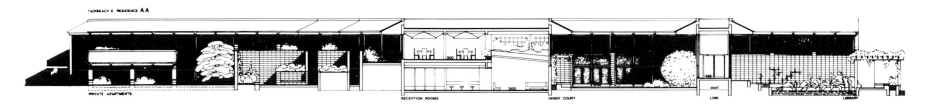

Taoiseach's Residence, section and first floor plan

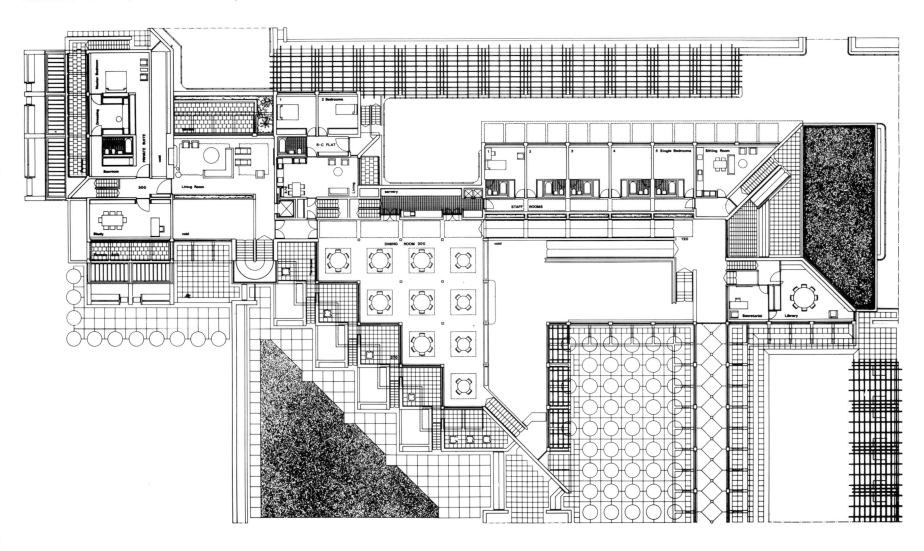

Norman FOSTER

1QRC – EXTRACTS FROM A PROJECT DIARY

1 QRC is the abbreviation used by the Hongkong and Shanghai Banking Corporation to describe the site of their headquarters in Hong Kong, otherwise known as 'One Queen's Road Central'. It is located at the head of Statue Square, which is the political and economic centre of this trading city and one of the few public open spaces in a densely populated settlement where nearly five million people, ninety-eight per cent of them Chinese, occupy some twenty-nine square miles; an area roughly the size of Guernsey or Manhattan. 1 QRC as an architectural project started formally in June 1979 with a telex from the Bank to seven firms of architects around the world. Extracts from this and various other documents follow to give some insights into the background of the project and expand the accompanying design drawings and photographs.

73205 HSBCD HX

To Norman Foster Foster Associates
12 Fitzroy St
London W1
From The Hongkong and Shanghai Banking Corp Hong Kong
13 June 79

From R.V. Munden

Request for Architects' Proposals

Dear Mr Foster

As you know we have recently been studying the redevelopment of our headquarters at 1 Queen's Road Central Hong Kong. The present building was constructed in 1935 and provides a space of about 300,000 sq ft. A completed redevelopment could amount to about 800,000 sq ft.

At a meeting of the Board in April it was decided that two types of phased redevelopment scheme should be studied further and that a decision on redevelopment should be made in the Autumn. The Board considered that the quality of the site and the importance of the project to the Bank justified selection of an architect on an international basis. An international architectural competition was considered, but rejected as impractical in the time available.

After consultation with the Royal Institute of British Architects and others the Board has decided to invite proposals from seven architectural practices of whom you are one. The terms of the Brief and accompanying information on the background of the project are being sent by courier to our London office at 99 Bishopsgate where you may collect them.

An initial discussion will be held on 11 July in Hong Kong and submissions of your proposals will be required by 6 October. Live presentations of your work may be required in early November, and it is intended that the Board will make a decision on future action by the end of November. In making this decision the Board will be assisted by Mr Gordon Graham, President of the RIBA as Architectural Adviser and other advisers on technical matters. Mr Graham has also assisted us in establishing the brief.

Sketch from the first visit to Hong Kong, July 1979

The principal aims of this three month exercise in order of importance are:

to help the Bank decide on an approach to solving the problems of which of the two types of
  scheme is the better;
to help the Bank decide whether this is better than doing nothing;
to select and appoint an architect.

With these aims in mind the Brief has been made as broad as possible. If you wish to be considered for this appointment we shall be pleased if one, two or three representatives of your practice including the architect who will be in charge of the project will visit us for the initial discussion on 11 July. At this discussion you will be briefed further on the background to the project and will be given the opportunity of asking questions. Further information on the initial discussion will be issued in due course.

Please confirm as soon as possible your acceptance of this invitation and your intention to attend on 11 July. Your reply should be addressed to N.A. Keith, The Hongkong and Shanghai Banking Corporation, 1 Queen's Road, Central, Hong Kong. Telex no 73205 HSBCD HX.

Yours sincerely
R.V. Munden
ENDSCN 1257
261571 FOSTER G
73205 HSBCD HX

The ensuing visit to Hong Kong opened up new experiences, both of the Far East and the business of banking. Impressions were vivid; the sheer vitality, energy, bustle, noise and congestion quite unlike any other urban experience. The alloted time passed quickly with briefing sessions, guided tours of the existing premises and the inevitable socializing when some twenty architects are brought together from around the world. After the formalities were over we tried to assess how to make best use of an extended stay. It seemed at the time that the key to the architecture lay in the integration of new-generation banking activities within the urban context of Hong Kong. It was therefore necessary to study how the Bank worked in greater depth and also to understand more about the site and its relationship to the total city. We posed endless questions to learn more about the commercial activities and spent long hours sketching, photographing, pacing and just plain watching the site.

We returned to London with a growing preconception that there was a more gentle option than the two site development alternatives so far identified by the Bank. This view was sustained during the following three months and became embodied in design proposals. These attempted a fresh look at the social and spatial implications of work and travel in a high-rise office tower, as well as the way in which such a building might knit into the urban fabric. By careful reference to the structural seams of the existing Bank it was shown that this concept could be effected in easy stages, to realize an integrated mix of old and new.

This direction, which we called 'phased regeneration', was clearly different from the Bank's stated brief. With no opportunity for a dialogue the challenge was to communicate the thinking at long range, using models, diagrams and drawings. The following extracts from the introduction to the competition report give a flavour of how the concept was broached.

The first recommendation to the Bank is that the criteria upon which the proposal is to be based should be expanded. Seven points are suggested ....
....

This new criteria provides a more rigorous yardstick against which to measure the Bank's two redevelopment strategies. They are examined and this process leads to a new concept called 'phased regeneration' ....
....

The proposal makes a firm architectural commitment within which it is possible to plan almost infinite internal variations – both now in the discussion stage and later during the life of the building. The ideal building would be rooted in such realities and be able to optimize on them ....
....

However, architecture is still about making appropriate spaces. From the outside the new building completes the space which is called Statue Square. Although appropriately larger than the existing, the new building would be in scale with the area and dominate it by presence rather than sheer height. The report only tells a part of the background. What it cannot show are the endless discarded models and drawings ....
....

Sketch study of the parts of the existing building
Original Bank building of 1935

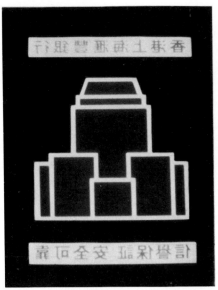

Hammersn

Willis Fabe

Sainsbury (

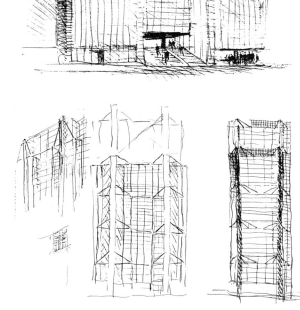

Sketches of the development of the building from the original competition entry through the 'chevron' and 'cluster of towers' schemes to the final version

But the Bank also has a symbolic presence because of its operational reality. This and the site formed the starting point of our studies in Hong Kong. New kinds of working spaces are suggested which could be operationally more efficient and socially more enjoyable ....
....

The building becomes an extension of the organization — reflexing to accommodate its demands rather than blocking them ....
....

The approach makes it possible to retain the existing Banking Hall virtually intact, almost to the end of a total site development. Alternatively, decisions would still be open on whether to retain it permanently, with or without the existing North Tower. If, as recommended, the Banking Hall is replaced, then the handover to a better and more appropriate space would be one clean move. The scheme which achieves these goals is called 'phased regeneration' because it is especially sensitive to the anatomy of the existing structures and views the bank operation as a totality — embracing new and old alike — complete at any stage, even if unfinished. This organic approach is the opposite of running conversions which are in a constant state of transition, part old and part new. The process eventually produces a new building uncompromisingly looking towards the future but inspired by the past.

Following their evaluations of all the submissions, the Bank invited us to return to Hong Kong for discussions. After three weeks intensive joint work a presentation was made to the Board, the outcome of which was the following press release in November:

The Hongkong and Shanghai Banking Corporation has decided to redevelop its headquarters building at 1 Queen's Road Central, Hong Kong.

The Bank's Board has commissioned Foster Associates as architects for the redevelopment, and design work will start immediately with the intention that construction will begin by the end of 1980.

Announcing the decision, the Bank's Chairman, Mr Michael Sandberg, said: 'A complete redevelopment of the Queen's Road site is an expression both of the Bank's commitment to Hong Kong and of our confidence in Hong Kong's future as an international financial centre. We believe our new headquarters will not only meet the group's needs for the foreseeable future but will also be an exciting building of which Hong Kong can be proud.'

In the following months design went through many stages of reappraisal, emerging with a more developed version of the original scheme in which clusters of office floors were suspended from towers of structure, lifts and services. Sandwiched between these clusters were generously proportioned reception spaces with outdoor gardens, in which movement would be transferred from high-speed one-stop lifts to local escalators. These reception spaces would be oasis-like buffers between the tempo of outside streets and the denser office areas which they served. Such a breakdown into smaller village-like units, with a considered sequence of spaces from street to workplace, was felt to be more humanistic than the traditional office tower — quite aside from a number of operational benefits. The exterior expression of these spaces provided a desirable breakdown of scale, and a welcome relief from the bland anonymity of most modern office facades. This gave rise to much debate with the Bank; mostly conversational, but sometimes in an exchange of letters, such as the following:

2 January 1980

Dear Norman

I had almost forgotten sending you a copy of the *Fortune* article. Your philosophical reply of 17 December is much appreciated, especially the sting in the tail.

I think it must be true to say that *all* human systems are fragile because of human imperfection. The degree of fragility will depend upon the kind of system and how much human intervention is required to make it work. If we use the word 'system' to describe something like banking, which consists mainly of an interrelationship between human beings, then we are speaking of something extremely fragile.

On the other hand, a building is a system which need not be fragile. A cave remains the most solid form of house because human beings do not have to build it. Stonehenge still performs its astronomical purpose accurately because its builders kept things simple.

But Stonehenge is not very flexible and nor are the greatest cathedrals of the world. Their builders did not concern themselves with doubts about the future. They assumed that the purpose for which they

were
shou

In s
unce

Ther
obvi
not

'Ser
the
ever

At n
that
rese
imp

One
by t

You

Roy

11 J

Dear

Than

If Si
and

Sadl
endu
finel
of g
that
is so
for s

Thir
trad
spiri

I car
virtu
part
phys
Fran
arch
buil
lectu

'Mo
Our
'hol
light
Brid
mac

'Our
Feuc
mas
ferro
obvi

although the technology of many of the units such as the engine and electronics have changed, and in some cases quite dramatically, it is still in full production. Even though its appearance is that of a fixed object, its design concept is based on modular units of limited life on a long-life airframe and it still responds, like most contemporary flying machines, to this process of continuous change. A word about the expression of change. The Bell 47 with its exposed truss structure, visibly bolted on appendages and giant bubble cockpit was, in many ways, a pioneering helicopter. Although its appearance was symbolic of infinite change it has ironically been overtaken by events and is now no longer in production. The aesthetic of the Bell 47 is that of seemingly *ad hoc* assemblage and it is in a tradition of other objects, such as American trucks with their overlays of highly customized variations, as well as such space hardware as the lunar modules and moon buggies. This aesthetic of infinite change does not always truly represent the realities of change. Conversely there are enough pertinent examples from that same American machine culture whose continued longevity belies their seemingly fixed and more integrated expression. For example the 'Airstream' caravan trailer, buses and aforementioned 'Jet Ranger' helicopter are, despite their fixed cocoon skins, firmly in that tradition in which renewable short-life items which can be easily retrofitted into a longer-life rack, chassis or airframe. The parallels in architecture are self-evident.

Our scheme for the Hammersmith Centre, which is a ring of fixed offices around a large-scale public space capped by a transparent bubble roof, is in principle a magnified diagram of the Willis Faber building. To us it also evokes references to a personal favourite, the Victorian Palm House at Kew. Again the detail planning, with service nodes at the entrance 'gateways' and variable infil on the public edges, allows much scope for detail change within a more fixed instructure and overall shell. If the shell of a flying machine is in part an optimum response to the external aerodynamic constraints then likewise the shell of this building attempts to respond to the environmental constraints, both quantitative and qualitative.

THERE IS A MAN-POWERED PLANE THAT WENT ACROSS THE CHANNEL, CALLED GOSSAMER ALBATROSS.

I think it was relevant that the designer, Paul McCready, achieved the first man-powered flight because, unlike everybody else, he did not try and copy what everybody else thought such a craft should look like. He went back to basics, the pioneering work of Lilienthal and Wright around the turn of the century, and combined these early principles with new high-performance materials such as carbon fibre and Mylar. He spoke to an audience of architects and designers at Aspen in 1980 who all thought that his design process was highly creative. Paul McCready did not agree; he felt that Kramer, who set the goal and prize for the first man-powered flight, was invoking the truly creative act. The issue of patronage has likewise always been central to the creation of pioneering architecture.

I would like to end up on flying because like architecture it is something of a personal addiction. I fly a glider because it gives me a lot of pleasure but it is interesting that with the passage of time gliders have become faster, they can cover far greater distances, they are much safer, they are more comfortable, they combine low energy and high technology and they give their users a great deal of pleasure. They are also the most incredibly beautiful objects, whether working or at rest.

I think there are some lessons here for architecture.[2]

Although the subject of 'technology' was raised in this lecture and the informal discussions which followed, it does seem to lead to much confusion, particularly with journalists who seize on the word quite out of context. Technology is surely only the means to an end, always has been from the first man-made objects onwards and cannot be seriously considered as an end in itself. These attitudes were perhaps made more clear in an interview with the film director John Read, who produced a BBC documentary arts film about our buildings in 1980.

JOHN READ

When Sir Robert and Lady Sainsbury decided to give their permanent collection of art to the University, they also agreed to pay for the building. They wanted it to have the informality of a home, where things of quality are lived with and loved.

They wanted something original and modern in spirit. It had to combine the direct experience of art with the teaching of its history. It also had to open up these pleasures to the whole university and to the general public as well.

I suggested to Foster that it was not easy to associate the technology of the building with its character as an art museum.

NORMAN FOSTER

I'd like to think there weren't preconceptions. And this word 'technological'; 'technology' as a word is very loaded. I'm told that a root word of 'technology' – *teknos* – originally meant 'art' and that

2 Extracts from a talk by Norman Foster at Centre Pompidou organized by *Architecture et Construction*, 26 February 1981.

127

the Ancient Greeks never actually separated art from manufacture so they never developed separate words for them. So really technology is about making things and works of art are about making things.

It is interesting to see the project through the eyes of others close to its realization and from different cultural backgrounds. Kenji Sugimura, a Japanese architect working on the project, wrote a lengthy article for the magazine *SD*. The following brief extracts give his personal insights into the design process and how he sees the complementary influences of traditional architecture and new materials.

INCORPORATING THE TRADITIONAL

For example, we are now experimenting with the application of Japanese *shoji*, as sliding screens, to the inside of glass walls in place of blinds. *Shoji* are, in effect, beautiful screens which can provide a 'cushion' against the excessively powerful summer sunlight of Hong Kong and yet are translucent, so that there is no undue loss of brightness. Norman Foster himself is a keen student of the sophisticated use which traditional Japanese architectural space makes of light and shade. Following the principle adopted by Foster Associates of having direct personal contact with the relevant materials in the course of their scientific architectural work, staff members have flown to Japan and looked for themselves at the ancient buildings of Kyoto. In addition, samples of actual *shoji* have been brought back to England, studies made of the physical properties such as U-value and light transmission rate when screens having a similar effect are made from glass panels, and repeated tests carried out of their environmental and visual applicability. Strange though it may sound, having come from Japan myself, I have been compelled, in our London design office and with a view to a building to be erected in Hong Kong, to take a serious new look at our Japanese *shoji*. It has not yet been decided in what form they will be installed in the Hongkong and Shanghai Bank. This concern with *shoji*, however, is of great interest as an example of an attempt to abstract a concept from traditional architecture and accommodate it to a high-technology building.

THE SEARCH FOR NEW MATERIALS

A central programme in our design work is the study by each individual architect of the fundamental properties of materials, at first hand. Our workplace is awash with samples not only of building materials but also of parts of industrial products from a variety of fields – the aircraft, automobile, and shipping industries, among others – collected from all over the world. We have to keep a particularly keen look-out for new materials in which improvements have been added to the original properties of the basic material. Realistic studies of space are often carried out, using thoroughgoing mock-up models. Several mock-up builders, employed exclusively by us, are constantly at work; in the fully equipped workshop, they give three-dimensional reality to our drawings, which are being modified almost every day. When I first came to this office and saw the constant success of metal-and-plastic mock-ups being produced, I was much impressed with their British crafts-manship.[3]

As the project gathers momentum both on site and through major subcontractors producing components and modular units as far apart as Japan, USA, Germany and the UK, the paperwork grows exponentially. Despite the application of computers, word processors, telex and facsimile links between London and Hong Kong there is seemingly no alternative at this point to the traditional despatch of documents and continuing preparation of reports. A single 'Architects Instruction' can typically generate in excess of 500 drawings for postal collection. A recent instruction was accompanied by the despatch of some 2,000 documents. The range of reports is wide from evaluations of competing bids through to options studies of internal transportation systems. Although time-consuming, the process of justification for approvals from the Bank is an essential part of the formalities which complement the more informal day-to-day contacts. The quantitative kind of report is relatively straightforward; the more subjective and qualitative sort is far more difficult as the following might show. These are concluding extracts from a report 'Colour in the Building' requested by the Bank and prepared in July 1982.

INTEGRATED APPROACH

It will become apparent that the issue of colour cannot be seen in isolation from other decisions which inform the total design; rather it is an integral part of the total that follows in the logical sequence of project development. This is not to imply that there are no subjective aspects to the selection process; such is clearly not the case. On the other hand in an integrated approach the choices do narrow down within a wider hierarchy of design decisions.

Mock-up of mullion sun visor and blinds

3 Extract from *SD* Magazine dated March 1982.

Manufacture and testing of cladding, Cupples, St Louis

EARLY SCHEMES

The first competition model showed a clear external expression of structure singled out in a metallic grey colour. It was shown in slightly differing versions to the Board in 1980 and 1981. Although the detail of massing and structure showed minor changes, the expression of a predominantly grey exterior has been constant in drawings, models and photographic montages. A carefully co-ordinated colour policy for the total building must obviously go deeper than such exterior images might infer and to communicate this it is first necessary to explain how the building is structured visually, both inside and out. It is this close visual relationship between exterior and interior that is unusual today and it could be argued is different from most other currently prevailing approaches to design.

STRUCTURE DEFINING SPACES

In the new Bank Headquarters the double-height spaces which occur at intervals up the building are an obvious example of the structure forming the spaces as well as providing a support from which the clusters of office floors below are suspended. What is less obvious, except from perspective drawings and models, is the manner in which primary and secondary beams in the suspended floor structures are revealed on the inside of the offices themselves particularly the ceiling soffits – a clear expression of the structural system which breaks down the scale of the spaces to a more human and intimate order. Given such a visual order which both links and divides up the inside and outside of the building with clarity and logic, it is an obvious follow-through to enhance that visual articulation of the frame and its infil, by a sensitive and consistent use of colour on a form-related basis. The new Bank Headquarters is in a clear tradition of framed buildings, the history of which provides ample

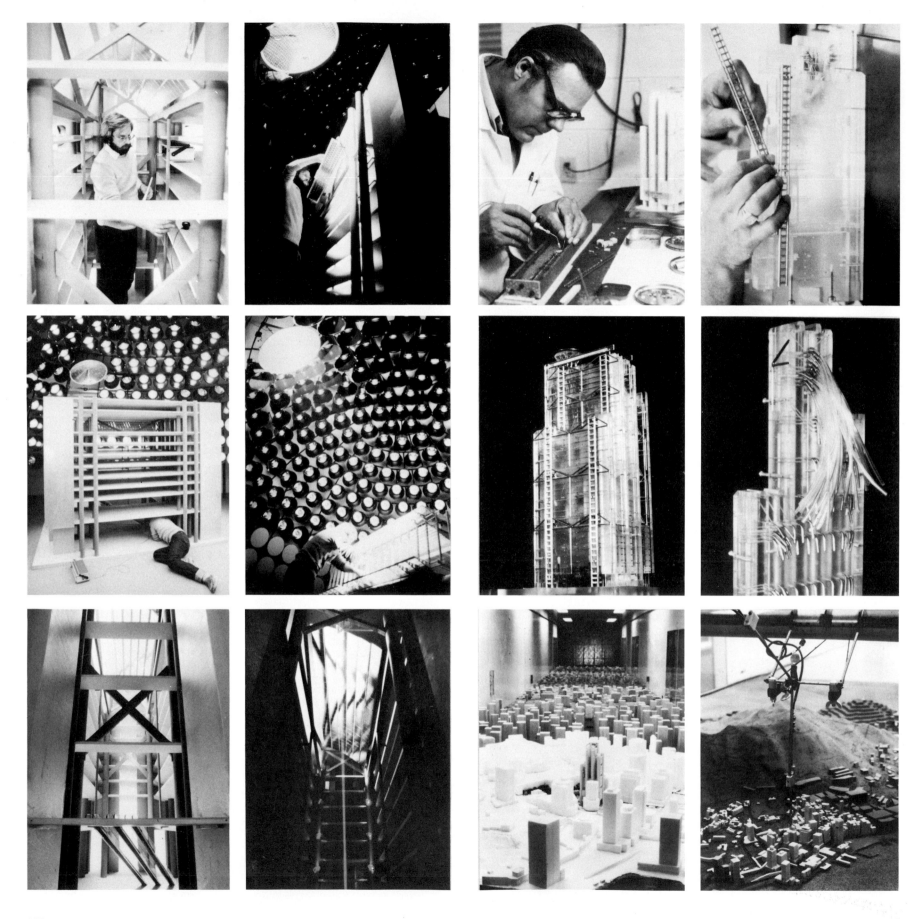

evidence of similar approaches. The interior and exterior spaces of Tudor architecture in Britain are as clearly labelled by coloured frames and contrasting infil as they are in the traditional Oriental architecture of China and Japan.

EXTERIOR

The consistent theme of a grey-coloured structure visible both inside and out provides a calm framework against which contrasts of colour and texture can be enjoyed. Externally these range from the reflective surfaces of glass windows with their fine tracery of silver mullions through to solid panels and grilles in a complementary variety of greys and silvers. Discrete accents of brighter colour are important in such an overall muted scheme and would come mostly from visible parts of the interior, as well as from external signs at the summit and base of the building. A major contrast of colour and texture is provided by the gardens which occur at intervals up the face of the building, both front and back. These are designed to overhang as a lush foil to the angular structures which enclose them; not only for aesthetic purposes but also to help achieve a sense of orientation inside the building and break down the scale on the outside. At night these would be flood-lit to emphasize them as hanging gardens in the sky as well as to provide important colour accents.

INTERIOR

Internally the approach to colour is the same; moveable partitions and screens express a grid of colour, probably using traditional Chinese reds and greens. The curved reflector panels which are recessed between the structure ceiling grid are of polished aluminium – sculptured to 'model' the interior spaces as well as providing the ultra-low-energy lighting system. The larger panels in the wall plane will, depending on their location, either match the external panels or be simple, white surfaces, patterned occasionally by perforations for acoustic control purposes. The plane of silver-grey carpet is broken by the junction lines between panels – a kind of twentieth century version of the Tatami Mat in traditional Japanese architecture. Although the scale and context are different the principles are similar. Both are rooted in the physical limitations of material, sizes, portability and a concern for an overall visual order.

GLASS AND MODELLING STUDIES

The extent and nature of the glazing has an important influence on the internal and external appearance. In view of the wide variety of spaces and activities within the building it was felt appropriate to avoid the simplistic images of one kind of glass throughout. The Banking Hall, for example, has main windows which stretch the full height of the ten-storey space on the east and west flanks of the site. These windows use a special sandwich of glass and insulation to ensure the right quality of natural light to create an inward-looking space around the atrium and avoid the prospect of dominating views of the adjacent Bank of China and Chartered Bank. The whitish soft translucent appearance is reminiscent of the light-diffusing screens in traditional Oriental architecture and is a newly developed sandwich of glass sheets and fibreglass lamination which we refer to as *shoji* glass. The walls to the north and south of the Banking Hall are clear to take advantage of the fine views out to the water and back to the greenery of Battery Path. It has been possible to go a long way towards simulating the appearance of the finished building. Such studies show that in spirit the facades are closer to the more richly modelled exteriors of classical buildings than the planar forms of 'traditional' modern buildings. This is partially due to the overall massing with its set-back but also to the modelling and filligree of detail that such devices as the sun breakers and grillages provide.

PRESENT-DAY COMPARISONS

The vital ingredient which informs the design of the new Bank and occurs in so many historical examples is that of an ordering geometry which can be seen both inside and outside the building. It is this sense of order which shapes and informs the spaces, breaks them down in scale, imparts a sense of orientation and in the end provides a special humanizing effect. The antithesis of this approach is a typical twentieth century commercial building. Its outside is physically and visually divorced from the inside, which is probably wall-to-wall anything that conveniently comes to hand from a catalogue – ceiling tiles by the millions, carpet or marble according to the interior decorator's whim or budget decree. It is hardly surprising that in large buildings, where the effects are magnified by scale, the occupants become disorientated. The bland anonymity of repetitive office floors is faceless, varied only by the numbers in the corridor or lift car. Outside the beholder is likewise intimidated and dwarfed by a mindless wall-to-wall curtain wall – the ultimate veneer of yet another dumb-box.

Several months have passed since that last report was drafted. The central ideas have so far survived intact despite, or maybe because of, the process of intense auditing and justifications. Much detail design work remains to ensure the successful realization of some of the more radical new concepts such as the sun scoop and cast glass floors to the public plaza. The pace of

ESCALATORS

EVERY ESCALATOR IN HONG KONG (AND EVERY ONE I SAW IN THIS RECENT VERY BRIEF TRIP TO JAPAN) HAD THE EDGES — PAINTED IN YELLOW.

ARE WE GOING TO BODGE OURS IN THE SAME WAY ? (I SUSPECT IT IS EITHER A LEGAL SAFETY REQUIREMENT - INSURANCE PROVISION - OR BASED ON ACCIDENT RECORDS)

WE MUST DESIGN OURS IN – TO BE SAFER & LOOK BETTER !

OSAKA 23-7

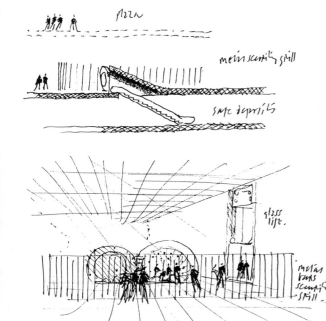

Notebook reminders

the programme is such that the service modules have already been finalized and approved in full-size finished state. These are the prefabricated elements, some 140 of them which will stack 180m high to produce the four towers of mechanical plant and toilets. They are already being manufactured in Ako City, Hyogo Province by a Japanese industrial consortium and will literally roll off the production line into a sea-going barge for transfer to a container ship at Kobe which in turn will transport them virtually direct to the Hong Kong Island site. Each module is 12m × 3.6m × 3.9m, virtually a building in itself. Almost everything within has been specially designed for the project – from partitions and floor materials through to wash basins, light fittings, door furniture and ash trays. The development process has involved three full-size stages of mock-up and prototypes, although the building site itself at this time is still only a hole in the ground. Other newer design challenges are still emerging. The base and summit of the building demand more work especially with the potential to incorporate large-scale air curtains to the two main road frontages and thereby climatically control the public plaza. This and the demands of security against typhoons and insurrection pose further challenges. How, for instance, can the large-scale grillages required to effect such protection be compatible with free public passage? Interestingly such elements also provide clues in other directions – they are yet another ingredient not only to introduce colour – Chinese Red – in the same spirit as the internal screens – but also to symbolize those other qualities special to a Bank – literally secure and seen to be secure, manifest like the structure – echoes of that early Board meeting 'our building must look like a Bank ... I wonder what they meant?'.

Right *Night-time montage*

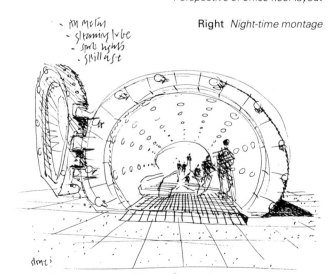

How to be secure and exploit the drama of seeing and experiencing security

Denys LASDUN

THE ARCHITECTURE OF URBAN LANDSCAPE

It goes without saying that the tradition into which an architect is born, is of primary significance in his subsequent development. The tradition in my case was the Modern Movement. This sprang from an artistic revolution of major importance. Cézanne was its chief progenitor. He looked beneath the surface of nature to the underlying structure and understood the paradox between the desire for total renewal in art and a deep attachment to the past. It was an attitude echoed by Paul Klee, one of the most perceptive teachers at the Bauhaus, dedicated to the idea of uniting art and industry, art and daily life. He too was as much at home in the remote past as in the present. He was able to recall a collective past without being imprisoned in it. He re-orientated tradition on to a new axis. Modern architecture thus retained a classical kernel. Functionalism, identified with the Bauhaus, was regarded not so much as a cardinal principle, but as a purifying agent in the architectural process. It demanded that architectural design be underpinned by reason and research in order to produce a sane and purposeful architecture. *For deep in every revolution, discreetly hidden*, Ozenfant wrote, *presides a classicism, which is a form of 'constant'*.

I remember as an architectural student in the 'thirties seeing these ideas embodied by Le Corbusier in a sublime building — the Pavillon Suisse. It struck me at the time that this building had the same bright morning quality as, say, the Pazzi Chapel — a seminal building which changed the whole course of architecture in its day. All the polemics of the machine aesthetic were brought together for a moment into tensed control and into taut and exact expression; the building became a laboratory of the Modern Movement, exploiting the Cubists' idea of interpenetration and free flow of space.

As important as a living architectural tradition is a profound understanding of the city — the characteristic physical and social unit of civilization. Cities possess size, density, grain, outline and pattern. People shape them and are shaped by them. Conflicting interests have to be reconciled; those least able to fend for themselves have to be protected. But our qualitative survival in cities depends on sensing the relationship of one part to another and to ourselves in time and space. We see old buildings as part of our mental landscape; we are disturbed and disoriented if this link is seriously ruptured. *If ever we are to have a time of architecture again*, Lethaby rightly said, *it must be founded on a love for the city. No planting down of a few costly buildings, ruling some straight streets, provision of fountains or setting up of stone or bronze dolls is enough without the enthusiasm for corporate life and common ceremonial. Every noble city has been a crystallization of the contentment, pride and order of the community.*

Royal College of Physicians, Regent's Park, 1960

Already by the 'fifties it was clear that the Modern Movement as a whole had failed to appreciate the needs of the city. The city was being gradually destroyed – its historical continuity was being lost, partly because of the needs and the complexity of modern life, because of the increasing number of people and traffic, but also because of the growing insensitivity to its requirements and the true scale of its activities. Cities were losing all those qualities which had previously been achieved by the slow organic growth which characterizes the living city. *The age of industrialism and democracy,* Henry Morris wrote, *has brought to an end most of the great cultural traditions of Europe and not least that of architecture... One result of this breakdown ... is to be seen in the disintegration of the visual environment in highly civilized countries in Europe with a long tradition of humanized landscape occupied by villages and towns of architectural character, sometimes of moving beauty.* He warned that the survival and revival of cities depended on recapturing the great tradition of European urban spaces which modern architecture and planning had almost obliterated and with it the most enjoyable parts of the city. *Modern architecture,* he wrote, *should not hesitate to use the geometrical forms that create the local precinct, the square, the three-sided court, the circle, the crescent, and in doing so it will not involve itself in the futility of imitation. These forms have a continuing social use and convenience as well as aesthetic influence. There may be,*

awaiting discovery, other forms of capturing and organizing space for the pleasure of man. It seemed probable that unless this warning was heeded, the arbitrary anti-urbanism of much modern architecture was going to prove fatal.

What modern architecture itself needed was some kind of re-orientation. I certainly felt myself firmly in the tradition descending from the pioneer masters of the Modern Movement. Early signposts of an even wider tradition to which I can still look back with confidence were Geoffrey Scott's *The Architecture of Humanism,* Ozenfant's *Foundations of Modern Art,* Rudolf Wittkower's *Architectural Principles in the Age of Humanism,* Summerson's *The Classical Language of Architecture,* the buildings of Nicholas Hawksmoor, an architect profoundly concerned with the roots of architecture and with the nature of space, whose works I measured and drew, Cubism, Le Corbusier's *Vers une Architecture* and his dictum that architecture is a thing of art, a phenomenon of the emotions, lying outside questions of construction and beyond them. But, for all this, I was fundamentally averse to Le Corbusier's Utopian idea, the Ville Radieuse, because of its lack of continuity with the long history of urban development and its failure to give due attention to diversity, to the handling of as many urban functions as possible in smaller areas free from traffic. I knew that there was no future in the sort of abstract urbanism which caused architects to work without a sense of place or of the past. I became interested in designing buildings which responded almost ecologically to unique and specific situations.

This attitude begins to manifest itself in the Royal College of Physicians[1] in Regent's Park which was built at the beginning of the 'sixties. It is a building that is central to our work since it tried to put back into architecture its links with the city while remaining within the tradition of modern architecture. It accorded with the Nash terraces without imitating them. Scale and materials generally matched and rhymed with the Nash buildings. But the important issue was the new building's disposition on the ground in relationship to those already there. Set at right angles to the road it formed, with the low-lying lecture hall, a court – defining a community of scholars. It seemed to me that spaces between buildings minister as much as any other quality to people's enjoyment and the well-being of the community. I became interested in the possible fusion of interior and exterior spaces. Now, twenty years later, this court is being reorganized as a medical precinct with the

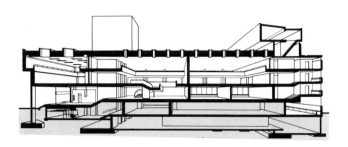

RCP – the city beyond

Nash terraces occupied by related institutes using the facilities of the parent building; and so this piece of city acquires a new life. Cities survive by self-renewal. Cities die without creative impulses in buildings.

The section of the building is concerned with the compression and expansion of space. You walk in and there is a lobby, then a larger reception area, a few steps and you finally come to an even larger and higher space which is at the heart of the building – the central hall and portrait gallery. When I once described this sequence of spaces at Harvard, they said, you take an awful long time to describe a classical hierarchy of spaces which simply goes, ah, ah-ha, AH HA! As you move up the square spiral free-standing staircase and look out on to the Nash terraces, the outside court and the inside space are fused into a single experience. This central hall (AH-HA) connects all the main rooms and encourages the College to do its work in a democratic, friendly and open way. Incidentally, the staircase, a dominant element in the hall, was originally written in the brief as 'the usual staircases, etc.'. As in Lutyens' buildings, the connecting spaces are often as important as the main rooms themselves.

The boundary between the inside and the outside has been obliterated and the interior and exterior spaces are fused together eliminating the 'facade'. In other words, the city beyond became the fourth wall of the central hall. I pursued this theme for the next decade, just when a new generation of architects was beginning to interest itself in the whole question of facades as an intervention between outside and inside, preferring more overt references to the past and more narrative content in the external expression of the building. This debate continues now in the 'eighties.

The world is certainly changing; the way architects work and the methods of putting buildings together are changing. Perhaps most important of all, the nature of patronage is changing – no longer having a passive role but actively participating in the process of architecture. At the same time, the self-confidence of the early days of the Modern Movement has gone. It is of course possible that all traditional values as we understand them could go overboard although major revolutions in architecture are rare. There are new perceptions but the world of architecture has become increasingly full of desperate and dangerous ideas and this is therefore no time for all-embracing nostrums. But somehow the architect has to find a bulwark of certainty on which his understanding can lean while his conception of his building as a whole takes shape. In a period of doubt, a wide variety of solutions is bound to be offered as part of the general search for certainty. That search is vital and every architect must take part in it. Some seek the

The theatre at Epidauros

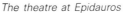

26 St James's Place, 1958
– Spencer House 1756

key in the recent past, in fashions that have only just had their day; others cast a little further back; others range right through to the Renaissance. These seek to re-establish the full Classical language, of columns and entablatures and assert that the archetypal buildings of the past can provide adequate guidance for present practice. They have chosen a hard and difficult road where success is problematic and failure probable. Perhaps it might be argued that there is nothing intrinsically wrong with an eclectic approach but the risks are great if it is not to degenerate into what Reyner Banham describes as just so much graffiti on the bodies of fairly routine modern buildings. Norman Shaw, the most refined and successful of eclectics, only just managed it because, beneath the ever-varying stylistic skin, he maintained an absolute mastery of the essential classical qualities.

Secure ground is certainly to be found in the relations of buildings to people and to cities and to the natural world, that is by getting back to what lies behind the rules, to classicism with a small 'c'; implying rational procedure, integrity of organization both inside and outside, respect for human scale, harmonious relationships, proportion, rhythm and repose. All this involves a disciplined humane architecture based on the wants and wishes of individuals. These qualities are the basis of all architecture. It does not have to be complex. A building like the Norman church at Kilpeck in Herefordshire can produce an unforgettable impression by its simple, sensitive and sensible response to place, weathering and to the unchanging human needs and desires.

To find a way through these problems we have to go back to the roots of architecture, to the forms of antiquity and their first full expression – and for me this means going back to the Greeks. They thought of their buildings in relation to the hills, the sea and the arch of the sky. The setting was as important as the building. The landscape – indeed the whole ambience – was inhabited by divine forces. At Epidauros, for instance, a splendid and exact geometrical composition is placed in a landscape that seems designed to receive it. Its radiant beauty springs from an elaborate system of responses, adjustments and correspondences in different parts of the auditorium. The setting-out point is a stone in the middle of the orchestra which has a radius the size of a Greek foot; this is the unit of measure used throughout. The radius of the orchestra is the module and it too controls the size and disposition of the whole design. Polycleitus, the architect, understood that if people in the upper tiers were to have a good view, the rake of the seating had to be increased. He also considered the spectators' comfort by hollowing out the bases of the seats to make more room for them to tuck their feet in. The fifty-five tiers of seats are arranged in two groups; twenty-one in the

upper and thirty-four in the lower. Twenty-one is the sum of the digits from one to six; thirty-four the sum from seven to ten. The design is clearly not an arbitrary one – it is the result of sober consideration, the consequence of the Greek preoccupation with numbers.

But the most subtle and significant adjustment of all is the increase of the radius of the theatre by three and a half metres for the last two wedges on either side. This adjustment opens up the amphitheatre so that it can embrace the landscape and so that the city of Epidauros could form a living backdrop, never absent from the thoughts of performers and audience – the City which to the Greeks was itself almost a divine thing, the only political organization that they were ever really prepared to tolerate. The landscape was in fact one of the elements, with the city and the architecture, out of which Polycleitus could construct the theatre and out of which he could conjure its effect.

His action for me reflects one of the most profound impulses in architecture – the sense of place. While the mode and scale of life in Epidauros were totally different from today, we nevertheless desire a similar quality of response to the situations we face and a similar demonstrable harmony of parts. We must aim to do what the great architects of the past would have done in our situation faced with our problems.

Architecture has to begin from some set of ideas, for us just as for Polycleitus. An architect needs to provide for himself the area of certainty within which he can work. This area of certainty gives a heightened meaning – a universalization – to the architect's personal attitudes and convictions. In this way he can produce for himself his own creative myth – a myth which is sufficiently objective and convincing to provide a world view and to give him a foundation on which to build. There must also be an element of subjectivity; the myth must be partly an expression of the architect's personality and partly of his time, partly a distillation of permanent truths and partly of the ephemerae of the particular moment. It cannot ignore the point which history has reached nor the point which architecture has reached, but it must be more than a simple reflexion of or reaction to contemporary social, political and aesthetic values, important as these may be. It can produce an architecture of logic and lucidity even in an age of intellectual sterility and civic decay. There are historical forces but we are not bound by them. Architecture cannot be a matter of reason alone, *Two excesses:* said Pascal, *excluding reason, accepting reason only.* The architect's aesthetic judgement is a function of myth and reason, a link between the emotional and the rational and the necessary means of entry into the realm of ideas.

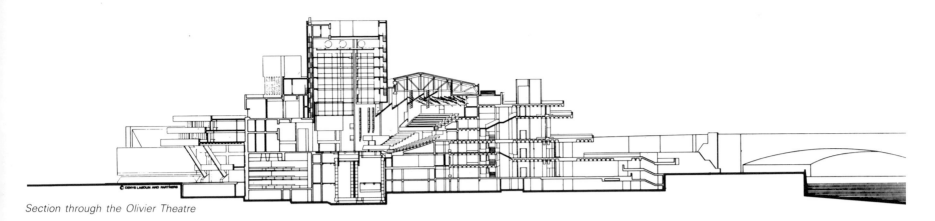

Section through the Olivier Theatre

Left *The National Theatre in relation to one of the most magnificent parts of London, the Kings Reach of the River Thames – architecture as an extension of the city.*

The National Theatre, Upper Ground, with IBM, 1979

My own myth is, I think, reasonable in quality. It engages with history; it concerns an architecture which is an extension of the city or landscape. It seeks to promote and extend human relationships. The buildings are related to other buildings which may be close in space, however far off in time, but they do not make stylistic concessions to the past. I look for a substructure from the past and try to transform this in modern terminology.

What I have been discussing has mainly concerned the interpretation of human relations through the medium of space and the fusion of interior spaces with the exterior grain of the urban landscape. But there also has to be a language which can express this inner core of meaning. The National Theatre illustrates one particular stage in this development. It is an architecture without facades, but with layers of building like geological strata connected in such a way that they flow into the surrounding riverscape and city. The building is thus an extension of the theatre into the everyday world from which it springs. The design attempted to offset current doubts about the permanent housing of institutionalized culture and form-making by responses away from the isolated monument to an architecture of urban landscape. The 'strata' are the basic vocabulary. Carefully proportioned planes are cantilevered from dark recesses of shadow and glass and are made to interlock interiors with the outer surroundings. Space is compressed and expanded according to need. The 'strata' can be used as roofs and walkways for planting. The soffits mediate as boundaries between inside and outside. The 'strata' express the visual organization of social spaces in geometrical terms. These spaces are not just the result of joining together functional requirements but are tangible elements in the architecture bringing order to the complexity of the programme. The 'strata' recall the streets and squares of the city and the contour lines of the hills. They bear witness to the roots of an architectural language inspired by natural geological forms. Architecture for me requires such a language of recognizable elements of classic simplicity, weight and interval; it must have a comprehensible and eloquent way of speaking so that its organization of space and modification of form can be easily understood and so that everyone can comprehend, appreciate and be moved by what the building is trying to do. This language must allow scope for intuition – the ability to grasp the significance of particular phenomena before the reason can begin to comprehend them, the instant apprehension of proportion by the eye alone unaided by calculation.

I have always looked on our buildings as actually or potentially part of something larger than themselves, as expressing or inviting the further development of the city or the natural ambience. **I never separated my ideas about architecture from those on the nature of cities which are there in time, have to change and have to grow; architecture is in a sense a microcosm of the city.** Cities are made up of relationships; building to building, building to street, space to space. The enjoyment of cities springs

The National Theatre, 1967

Left *Model of the abandoned National Theatre and Opera House scheme, 1963–65, on a site between County Hall and Hungerford Bridge – an architecture of urban landscape, of foothills and terraces descending down to the river and reconciling the Shell Tower with the riverscape.*

Right *The National Theatre and IBM London Marketing Centre, 1979. A fulfilment of the idea of buildings as a metaphor for landscape first expressed in the abandoned National Theatre and Opera House scheme, twenty years ago.*
1 Somerset House 2 Waterloo Bridge
3 The National Theatre
4 IBM Marketing Centre

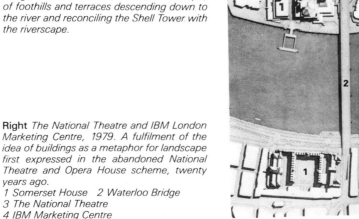

The meeting of two important buildings each with its own identity, conceived as a single composition in an amplified setting to match the scale of Somerset House across the river. The components of the architectural composition are the bridge, the waterfront, a reorganized traffic-free mall between the two buildings and a potential Theatre Square which would mark the point of arrival for NT/IBM and would be linked to a pier by the mall.

mainly from these relationships and connections, from qualities of the spaces between buildings, from parks, rivers, paths, alleys, roads, malls, crescents or squares; places, in other words, for comprehending the landmarks and organization of the city and its focal points of memory.

Architecture itself is expressed primarily through the medium of space. Space is the most elusive aspect of architecture, but it is its essence and the ultimate destination to which architecture has to address itself. The human significance of space, as we move through it, is that – like language – it is to be delighted in; it can be formal, informal, warm, cold, public, private, high or low in status. Space is perceived not by the eye alone but by all the senses. Sound in a building allows you to feel the acoustic dimension, the depth and resonance of the surrounding space. Space depends on light and light in turn collaborates with space to give it architectural form, tranquillizing or activating it. It is form that is life-enhancing. Materials and detailing also express the inner pulse and rhythm of the design. Architecture requires an appropriate sense of scale which relates to the human frame and which is, in turn, related to the larger order of the unity of the building. This unity relates to the even larger contextual order in which the building finds itself. *A house is nothing other than a small city,* Palladio rightly observed.

To me a building has to be positively linked to the city; it can be done by a harmony of facades, by making the building contribute to the street's language no matter what it is there to do itself. But the facade, although a magnificent device, is not the only one. Obliterating the sharp distinctions between interior and exterior is an alternative – the use of the City itself as a fourth wall: for example, Somerset House seen from the National Theatre.

Since architecture is overwhelmingly a social art and depends on the specific situation of the specific building, the problems of site, use, finance and situation have to be thoroughly considered before the mythic impetus, which will create the building, can be imparted. In each case, unique problems must be resolved on the basis of careful research. There are no inevitable solutions in architecture; a building springs from the architect's resolution, in collaboration with the client, of the individual problems in terms of the architectural language he has developed.

The client's role is vital; he must combine with the architect to lay bare all the essentials of the joint undertaking and give him the information he requires in a usable form by the clarification of the client's own ideas about the nature and purpose of the building he is commissioning and what it is to exemplify. These ideas must include adequate provision for the rights of others, particularly of other buildings. The process of architecture is thus one of continual co-operation, of study and discussion of present needs and future possibilities, of present desires and future hopes. This exchange of information is the life-blood of architecture; it will allow the architect to give the client not what he wants but what he never dreamed he wanted; something, however, which when he gets it he will recognize as what he wanted all the time.

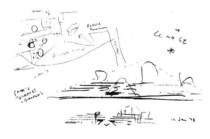

In this process models have a major part to play. William Curtis writes: *Models have a social function as a basis for dialogue. Most laymen cannot read plans but can understand a model which stands in three dimensions. The model serves to embody a visual hypothesis based on available evidence to date. It can stimulate the client to examine his assumptions and clarify his needs through a sense of the available alternatives, for it is rare that either client or architect know exactly what is required in advance.*

The dialogue is also a matter for participation and collaboration among the architectural team. It is for them to create order out of all the undifferentiated data arising out of the brief. They have to function not unlike a jazz group made up of strong individualists sharing a common language and grammar but free to express themselves within the total conception of the work as soloists or section members as the artistic situation demands. The creative process calls on that immeasurable quality which is beyond argument. Architecture is unequivocally an art. This implies giving form; touching the spirit; bearing the imprint of sensibility, temperament and intellect; concern for the immeasurable; the capacity to make value judgements as to what is ugly, what is beautiful; organizing plans that are in touch with life. All this is recognized by architects working together on a problem. Its presence in a building is sensed quickly enough by the perceptive looker and the ordinary user alike.

You cannot serve human needs, in terms of architecture, without a sense of form and space and, in my view, you cannot have form in architecture which is unrelated to human needs. Architecture only makes sense as the promoter and extender of human relations, it has to communicate through the language of form and space – through the fundamental classical qualities. *An architect ought to be jealous of novelties*, Wren said, *in which fancy blinds the judgement; and to think his judges as well those that are to live five centuries after him, as those of his own time. That which is commendable now for novelty, will not be a new invention to posterity. . . ; but the glory of that which is good in itself is eternal.*

Even in a fissiparous, disconnected age such as ours, where the volume of critical opinion greatly exceeds its content, it is inevitable that architects should be engaging in a search for symbols of continuity while being mindful of change. The art historical question is clearly not for me – I can only try to explain why our buildings are the way they are. It is true that the Modern Movement is coming to be recognized simply as one of the many manners of building in the long history of architecture and one that will one day come to an end like all its predecessors. But that day has not come yet; only tomorrow's historians will be able to tell whether it has died or not. What is certain is that it has accommodated many masters – Le Corbusier, Mies van der Rohe, Alvar Aalto, Frank Lloyd Wright, Gunnar Asplund and others who have between them produced sublime works of architecture, comparable to the masterpieces of any age. These are now lumped together with all the visually dull, repetitious, devalued and debased examples of 'modernism' which have been built in the last fifty years. This resulting confusion has obscured the fact that modern architecture is still evolving and still has the power to produce buildings of the greatest range and meaning without turning its vision backwards.

There are good and bad buildings in every period. It is their fundamental quality which will govern their final reputation. The essential will last, the fashionable will not. Architecture is the crowning art which defines and defends all civilization. That is why it must take a comprehensive view. Architects bear a unique responsibility as masters of the only art that must constantly affect people's lives. In fact, architecture is more than an art and more than a science; it takes in the whole of human experience. Only the architect has the chance to use the most recent technological developments in the service of a discipline that was already ancient when the Pyramids were built. The architect is the direct and continuous link between the life of home and market place and the most esoteric fastnesses of art. As Henri Focillon said: *Architecture has to be subject to the needs of society, rich or poor; faithful to the building programme and climate; it answers collective needs even in the construction of private dwellings; it satisfies old needs and begets new ones; it invents a world of its own.*

1 Pevsner in a BBC broadcast in 1966 referred to this building as *post-modern* – a successor to the international style of the 'thirties.

The strata inside and outside are the basic vocabulary. They seem to capture the fundamental sense of theatre as a place of gathering and they provide a framework for the experience of visiting the theatre which takes the city itself as its backdrop.

We searched for a single room embodying stage and auditorium whose spatial configuration, above all else, would promote a dynamic and emotional relationship between audience and actor – between a fixed architectonic geometry of vision, acoustics and concentration and the chance irregular demands of dramatic performance. We searched for an open relationship that looked back to the Greeks and Elizabethans and, at the same time, looked forward to a contemporary view of society in which all could have a fair chance to see, hear and share the collective experience of exploring human truths. The room thus offers many possibilities and certain contradictions.

At the South Bank Board meeting, October 1966, Lord Olivier said that this auditorium and stage answer everything they have ever asked for in terms of an 'open' solution. It had perfect sightlines; an acting area of the right size; the possibility of having enthusiastic members of the audience close to the stage; the possibility of modifying the configuration of the seating against the stage to amplify the thrust or take a more bland view. He said that it had marvellous atmosphere; he particularly liked the raised banks of seating at the sides. In fact all members of the Committee were immensely happy with this concept.

Form, space, structure and surface made manifest by the nature of concrete. People and events will be the decoration. The spaces are not the result of just joining together functional requirements but are tangible elements giving order and enjoyment to the building.

TWO COMMENTS

To have launched a building as momentous as the National Theatre should have cheered us up, in a period of national blues. Instead we are told that we can't afford it, don't need it, and that Peter Hall is impossible anyway. These reactions, at first so depressingly suggestive of total loss of nerve are, in fact, the routine ones that greet all new national landmarks. They almost inevitably cost more and take longer to build than they were meant to, offend numerous vested interests, and are stylistically out of fashion by the time they are finished: Wren was sacked before St Paul's was completed, and smart young Palladians thought its architecture a disgrace to the metropolis. The building of the Houses of Parliament was punctuated by interminable rows, mainly because it cost too much and because the expensively up-to-date ventilation system never worked; well before it was finished its Perpendicular Gothic styling was considered effeminate and hopelessly out of date. The Law Courts had such a rough passage that they killed poor George Edmund Street. Like them, the National Theatre can afford to wait.

Mark Girouard
The Architectural Review, January 1977

I think our aims and objects in the National Theatre were summed up by a Japanese gentleman in a letter to Dame Frances Yates, in spite of or perhaps because of his lack of total grasp of the smooth niceties of modern English usage:
'Somebody like a Japanese may more easily understand the National Theatre perceptively or intuitively because it has something resembling the Noh Theatre in its ambivalence of intimacy and severity. In its intimacy of raw humble features and forms, based on the river bank, evoking hillsides of ancient Greece, it can most naturally allure and absorb every class of people into it. In its severity of abstract geometric forms and spaces, indicating the numinous of the primitive people, those entering unexpectedly encounter the time and space of a kind of ritual. They gaily enter the theatre and go out transformed in serious terms.'

144

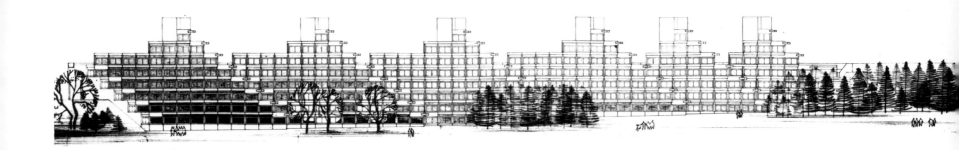

EAST ANGLIA UNIVERSITY, 1962–68

I never separated my ideas about architecture from those on the nature of cities which are there in time, have to change and have to grow; architecture is in a sense a microcosm of the city.

An academic city on a site which rises seventy feet from the River Yare on a south-facing slope two miles west of Norwich. The Development Plan is like a map of human relationships in Rashdall's sense – relationships between teacher and teacher, teacher and student, student and student.

The buildings are conceived as architectural hills and valleys. From the air, they are like an outcrop of stone. From the ground, they hug the landscape which is itself preserved by the plan's compactness and remains distinct from the urbanity of the University. The students' rooms are in groups of

Grassed 'harbour' between the two groups of student residences

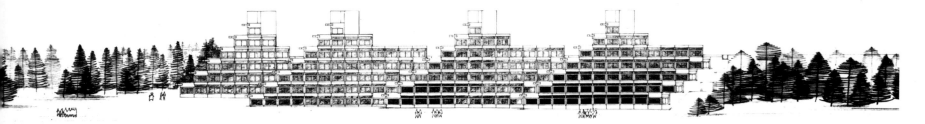

South-east elevation of student residences

twelve; they face outwards from the University to the south and descend in terraces down the slope towards the river and the lake. The rooms are linked by galleries to the centre of the University – the shape of the site allows access paths to enter the buildings at different floor levels; there are no lifts. The teaching and research accommodation is contained in a continuous belt of building which provides, in any part of its length, the various categories of space required by any school of study. This creates an architecture of urban landscape rather than isolated campus buildings.

The University from the River Yare

The University from the south-east with staff houses

Plan at elevated walkway level
1 Teaching Wall
2 Student Residences
3 Staff Houses
4 Library
5 Lecture Theatres
6 University House and Senate House

UNIVERSITY OF LONDON EXTENSION, 1965–78

The aim was to create on Bedford Way an ordinate piece
of London street architecture and, on the other side, to
maintain the scale and the areas of calm which had been
provided in the past by the Bloomsbury squares into
which the building of the Institute of Education penetrates. The wings descend in terraces towards the new
building for the School of Oriental and African Studies –
which is integral to the design – to make with it an
academic piazza. To date, only one wing has been built.
In 1978, a new Courtauld Institute and Art Gallery was
designed to take its place in the precinct. This was
probably the last chance to regenerate this part of
Bloomsbury – to make it as vital today as it had been in
the nineteenth century. Although welcomed by the
Royal Fine Art Commission, the GLC's Surveyor of
Historic Buildings and Sir John Summerson, the GLC
Committee did not support the application to remove the
remaining terrace fragments which was necessary to
permit construction.

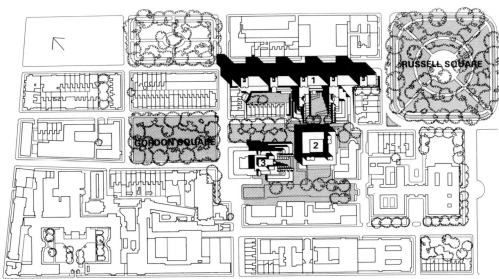

Site plan.
1. *Institute of Education*
2. *School of Oriental and African Studies*
3. *New Courtauld Institute and Gallery –
   abandoned project*

The wing descending into the precinct

The building in Bedford Way

RESIDENCES AT CHRIST'S COLLEGE, CAMBRIDGE, 1966

The principle of interlocking spaces, which formed the basis of the student residences at the University of East Anglia, is here developed further to create, on a restricted urban site, a low-rise architecture of urban landscape, in scale with the existing college despite the high density and diverse nature of the accommodation provided.

The form of the building, generated from the sectional organization, rises in a series of stratified terraces which is interrupted at the second floor level by a landscaped garden, grassed and linked at ground level to the court by a broad flight of steps.

The first stage has been built, the building has yet to be completed.

Detail of King Street elevation

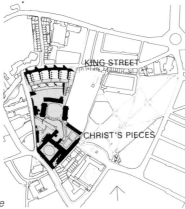

Site plan of complete scheme

Stratified terraces

THE EUROPEAN INVESTMENT BANK, LUXEMBOURG, 1975

The European Community's bank for long-term financing was founded by the Treaty of Rome in 1958 to counter poverty and disparity between countries and regions. It is sited on the Plateau Kirchberg, outside Luxembourg's city walls, on the great Jurassic ridge at the junction of two ravines sloping down to the river valley, as an essential element of the proposed new European capital. The intention was that the splendid situation of the building should be enjoyed to the utmost by those working in it. Air-conditioning was only to be provided in those areas whose use or position in the overall plan required it. The Bank thought that openable windows and close contact with the landscape would assist energy saving. The cruciform plan establishes the presence of the building on all sides; it occupies the site without destroying it. The landscape penetrates right into its centre for the enjoyment of all who work there. The use of materials, the juxtaposition of concrete and landscape and the geometrical ordering and articulation of elements is intended to create an ambience which while drawing little from classical academic practice reveal latent classical qualities of interval, proportion, rhythm and repose.

The building overlooking the ravine

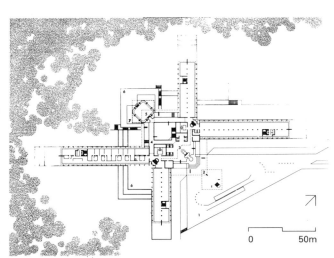

Plan at entrance level
1 Forecourt with parking under
2 Porte cochère
3 Entrance hall
4 Conference foyer gallery level
5 Terrace meeting room
6 Terrace
7 Water Garden

Terrace meeting room and water garden

The office module

The entrance quadrant from the forecourt

Conference quadrant terraces over the water garden

Section through the City

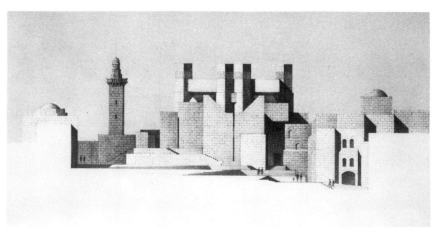

East elevation

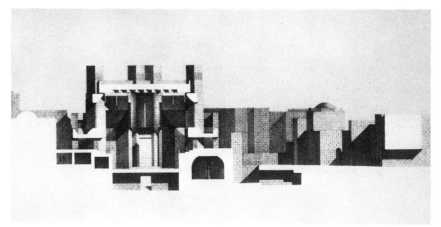

Cross section looking east

THE HURVA SYNAGOGUE, JERUSALEM, 1981

The site of the proposed rebuilding of the Hurva Synagogue is on the western boundary of the Jewish Quarter in the Old City of Jerusalem. A synagogue (twice destroyed) has been on the site since the thirteenth century AD. The Ramban Synagogue, dating from the tenth century, abuts at a lower level and stands adjacent to the Sidna Omar Mosque to the south. These two buildings generate the geometry of the sunken court which leads east and north terminating at the entrance to the Community Centre. Thus the literal and symbolic foundation of the proposed new Hurva is an historical fragment loaded with cultural memories and embedded in the floor of the City.

Four pairs of towers on each side of the building flank the central space and stand sentinel over it. Their symmetrical disposition makes it clear, on the outside, that the principal space lies between them and is felt to be a piece of the City set aside for higher purposes. Enclosure is completed by four apses, the men's entrance porch, two smaller women's entrance porches to the north and south, and the Sanctuary to the east which diffuses light from

behind the Ark wall. The dense thermal mass of the roof-strata hovering above the Assembly protects it against glare and heat and stands as a parasol above the events within. The cool night breezes of Jerusalem's microclimate

154

Detailed cross section through apses

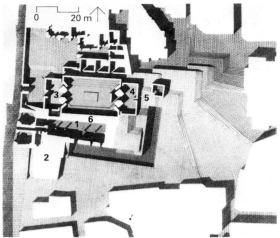

Plan at men's entrance level

1 Ramban
2 Mosque
3 Men's entrance porch
4 Sanctuary
5 Entrance to Community Centre
6 Women's entrance porch and gallery above

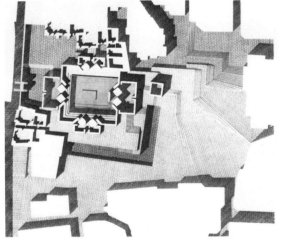

Plan at women's entrance level

are drawn down the towers by natural convection simultaneously releasing the heat from the building fabric. Walls are of stone, incorporating old existing stones. The roof is made of heavy pre-cast concrete sections with external ceramic fascias and the soffit is coffered. Windows are set in deep slots with the reversed reveals concealing the frames from the outside. Light, texture and the acoustical dimension of space combine to stimulate an imaginative response from the worshipper and reflect the inner pulse and rhythms of the design. Emphasis is on smallness and simplicity. The essential idea is that of the assembled community (300 men and 150 women).

The design may be thought of as a formally intensified abstraction of the flat roofs, tiny alleys and small squares characteristic of the ancient City. From close-to the Hurva is seen to be rising between and out of other buildings – an architectural mass, a unity, emerging from the diversity and grain of the City. It is not an isolated monument but an extension of the City itself. The Piazza is envisaged as a marketplace.

Because religious buildings are not unduly constrained by complicated functional demands they have given architects throughout history the opportunity to explore the human significance of space in response to a universal idea. There is no distinct traditional type for the synagogue. The formal vocabulary and the underlying themes of the design of the Hurva have evolved over the years from previous buildings, notably the Royal College of Physicians, the National Theatre and the European Investment Bank – all involved with the interpretation of human relationships through the medium of space and the fusion of interior sequences with the grain of urban surroundings. The Hurva continues these themes because they are felt to be appropriate and seem to bear directly on the core issues of religious meaning, just as Palladio's churches not only share features with his villas but are also an embellishment of the city.

To the question 'Does it look like a religious building?' the answer, today, can only be that it does *not not* look like a religious building.

THE CARLO FELICE OPERA HOUSE, GENOA: COMPETITION, 1982–83

The original Opera House built by Carlo Barabino and inaugurated in 1828, was bombed in the Second World War leaving only the hexastyle Doric Pronaos.

When an old building has in the course of time suffered minor damage, it may be best to restore it but, if it has largely disappeared, it has to be recreated from the new circumstances that it now has to face. No doubt an effective restoration of Old St Peter's would have been well within the range of Bramante and Michelangelo, but can we really feel that it was wrong of Pope Julius II to decide to start again and build it in the language of his own age? And this is the true spirit of Genoa – a City which has always been in the forefront of architectural development.

The Opera House demands something more than a pale copy of the past or a thin distillation of a few meagre 'primary elements'. It needs to express the complexities of modern ideas and modern living in ways that have been found acceptable to the twentieth century. It needs to draw on a wide vocabulary of forms which are both archetypal and contemporary and to use these forms in the service of a fully thought-out and integrated expression of the Opera House's functions.

The Pronaos is restored and remains the unchallenged focal point of the Piazza de Ferrari. It acquires a new independence and functional role as the Gateway to a Public Concourse leading to the Galleria Mazzini which terminates in the Largo Lanfranco.

The Belvedere expresses the pivotal point between the axes of the Pronaos and the Galleria Mazzini. By day, the form of the Belvedere matches and rhymes with silhouettes of the existing skyline. Views of the City can be enjoyed by the Opera-goers and at night the Belvedere will be floodlit.

Three positions for sculpture, suggested by Barabino (but not executed), could be incorporated as an integral part of the design proposals.

By placing the Opera House itself on an axis at right angles to the Galleria Mazzini, but still within Barabino's original rectangle, a generous space is made available for the creation of a Public Concourse linking Piazza de Ferrari

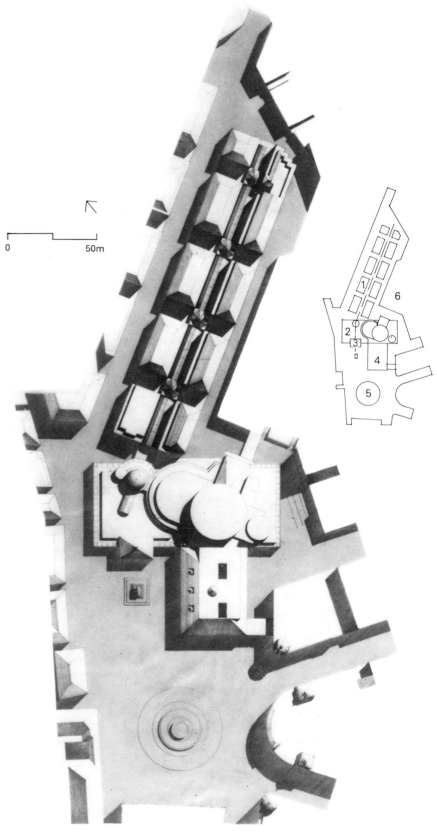

1 Galleria Mazzini 2 The remains of Barabino's Opera House, 1828, destroyed in the last war 3 Hexastyle Doric Pronaos 4 Accademia 5 Piazza de Ferrari 6 Piazza Piccapietra

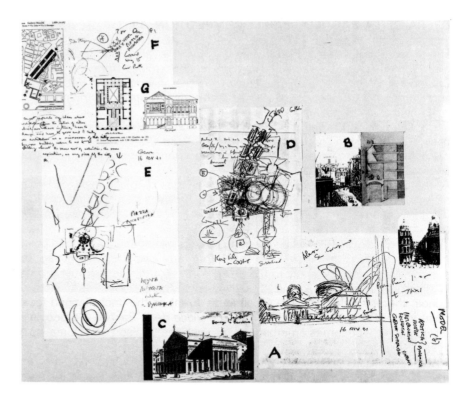

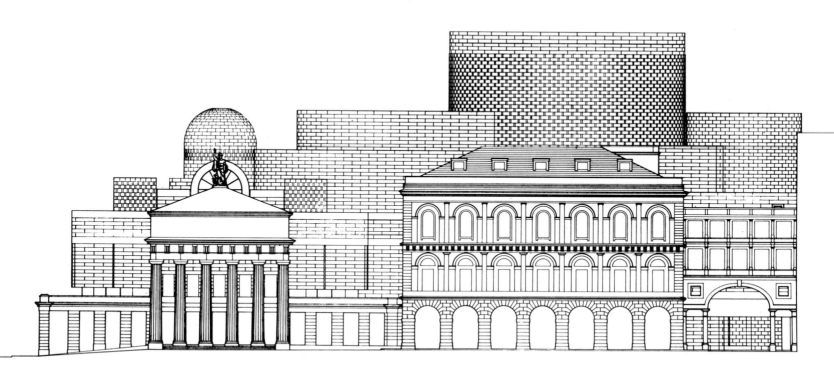

with the Galleria Mazzini. The two geometries, old and new, come together in a dynamic relationship which is reflected in the architecture. Furthermore the mass and configuration of the side stage makes a positive gesture towards Piazza Piccapietra. Finally, the flytower, echoing other circular forms to be found in Genoa, reconciles the axes of the Galleria Mazzini, the Opera House and the Accademia.

The Public Concourse in the form of an extended *galleria* is defined by the interior wall of the Pronaos, the glazed foyers of the Opera House and the circular walls of the auditorium. The Public Concourse acts as a converging focus producing a lively concentration of spaces available for shops, kiosks, events and 'happenings'. The entrances to both the Opera House and the smaller hall are from the Public Concourse which is overlooked by the private enclosed foyers of the Opera House. The circular wall of the auditorium encourages an easy flow of movement to the Accademia.

157

Leslie MARTIN

THE DEVELOPMENT OF A THEME[1]

Marcel Breuer once made the comment that the new architectural ideas of the 'twenties and 'thirties would have existed even without the advent of a new technology.[2] When that remark comes, as it does, from the inventor of a pioneering range of steel and plywood furniture, which is itself dependent on a new technology, it may be surprising, but it has to be taken seriously. Breuer was not saying that the new advances in technology did not exist. Nor was he saying that the range of the new architecture had not been broadened and expanded by the need to solve a whole series of new problems. What Breuer meant, I think, was that the overriding change came from the new way in which architects thought about architecture and ordered and controlled its forms.

There was, without question, a new freedom in the way in which buildings were composed around different problems and the ways in which space and enclosure were used by architects. And as a direct result of an open-ended method of design, the architects themselves continually widened and extended the range of ideas. In contrast to the current slogans and catchphrases, and the manifestoes which often seemed to suggest that everything could be solved by rational analysis and improved technical production, the more creative architects in each country continued to expand and widen their range of forms and had no hesitation in using the materials that seemed appropriate to a particular problem: sometimes it was an advanced technology and sometimes more traditional methods.

What seems to me to be impressive is the way in which in the last forty years or so the range of formal solutions has been widened. Each new generation has made its contribution and has added to and enriched that range. There are many lines of thought that individual designers have followed with sincerity and conviction. There have been buildings that have added significance and a 'sense of place' to parts of a city and, at the other end of the scale, others which although small and anonymous have made their contribution to the creation of a better environment. And in addition, a developing technology has frequently been the basis of imaginative and creative design.

All these things seem to me to be complementary and not opposing developments. Each separate line of thought will certainly be pursued with conviction and single-mindedness by individual architects. But when the history of the architecture of the last fifty years can be seen as a whole, these developments may appear as different currents within the broad stream of the evolving architectural thought of our time. What remains common to the kind of architecture that I am describing is that it operates through the *central medium of formal ideas built up around problems that have to be solved*. It is not concerned with 'fashion'. It adds to that growing and inherited vocabulary of ideas, forms and techniques that are appropriate to each real architectural problem: and it is this that in the long run is leading to the development of a new tradition.[3]

The work of my own studio can now be examined as it has developed over a considerable number of years. In retrospect it becomes clear that over this period of time we have worked on a number of different types of building each of which can now be seen as a development of one particular theme or line of thought. The types include university residential buildings, libraries, auditoria and art galleries; most of these modest in size and built within a limited cost. Each one of these categories now seems to have been a kind of exploration of the formal possibilities that arise from the study of the type form itself and the development of a set of ideas worked out within the limited constructional range that we were prepared to accept.

We began to work twenty-five years ago on university residential buildings. We started from the point that a form which had been known to work well for a community, the traditional College court, was worth using again. We followed this pattern in several buildings, varying the staircase arrangements and room clusters and then extended these ideas to suit linear and contoured sites. During the same period we were building a group of libraries in which we studied the generating form that seemed to work for libraries of various types. That particular study began twenty-five years ago: we are at the moment building another library in which the basic pattern that we explored still seems to be valid and is being re-adapted to meet the different needs of a music school library and the special conditions of a different site.

But for the purpose of illustrating the general thematic development of ideas, I wish to select from a range of designs for auditoria and in particular two schemes which seem to illustrate this continuity of thought. One is the Cambridge Music School designed some years ago; the other is a scheme for the Royal Scottish Academy of Music and Drama which is at present in the working drawing stage.

These two buildings have an ancestry. In the 'sixties we built an auditorium for the University of Hull in which the main hall, in this case used principally for speech, was sunk into a plinth of surrounding rooms. This arrangement provided two levels of entry to the sloping auditorium floor: one at the rear and upper level which provided the main public access from the entrance hall, the other at the lower level of the front seats. This low-level point of access provided the connection to a low-level promenade within the surrounding plinth which in turn linked together a series of rooms and spaces: for instance exhibition space, an art gallery and a chapel with a private court.

The programme for the Cambridge Music School again required an auditorium for an audience of about 500. In this case the principal requirement was a room with acoustics that would be suitable for the performance of music of various kinds. The hall was also to be used for teaching and for the presentation of opera, though it was accepted that extensive scenery would not be used and that a special type of presentation would be necessary. The supporting accommodation took the form of foyers, common room spaces, seminar rooms and lecture rooms (including one to seat seventy-five), an old instrument room to house an historic collection and to be used for teaching, a limited number of practice rooms and some backstage accommodation. A library connected with the main group of buildings was also required.

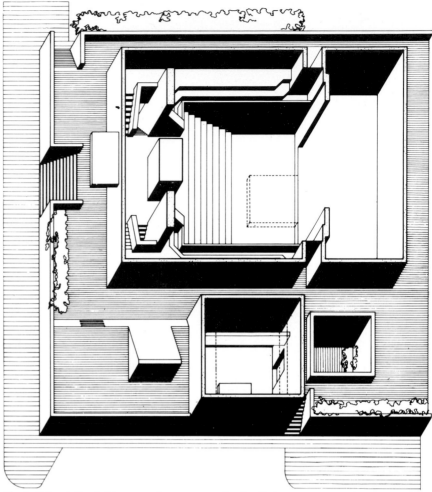

Music School, Cambridge opposite page *exterior detail*, above *axonometric*

University of Hull axonometric of the auditorium

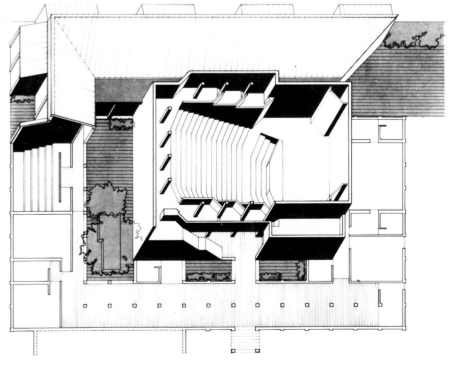

161

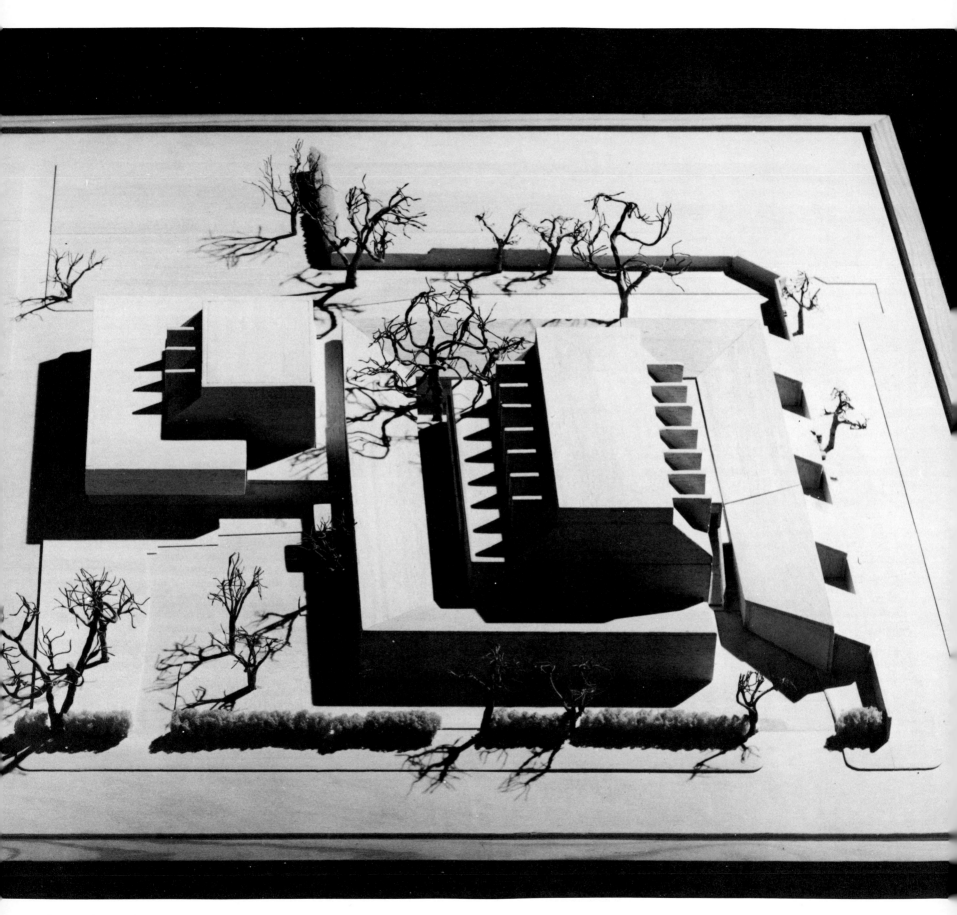

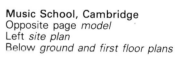

Music School, Cambridge
Opposite page *model*
Left *site plan*
Below *ground and first floor plans*

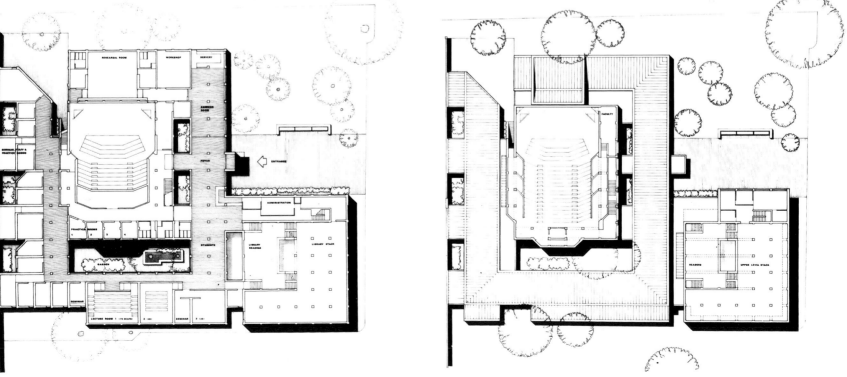

gai
the

rel
sys
circ
loa
red
tior
zin
this

see
des
nec
wa
and

Music School, Cambridge

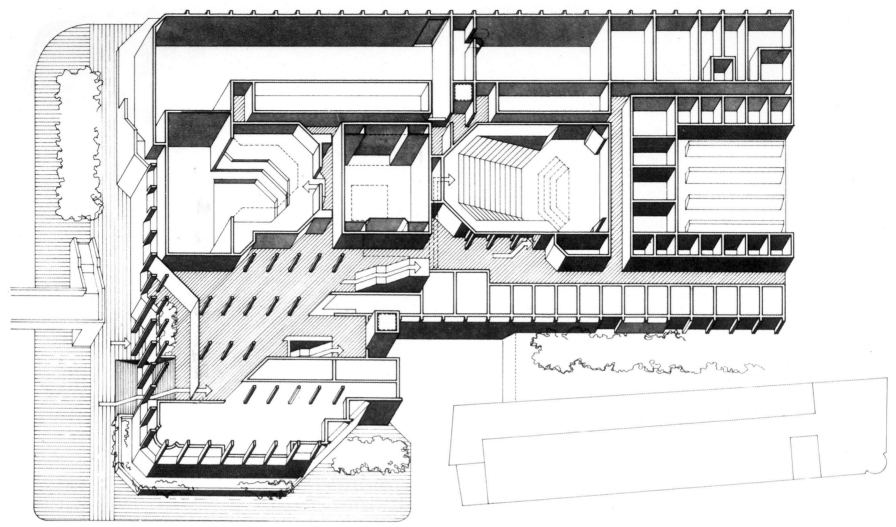

Royal Scottish Academy of Music and Drama, Glasgow *axonometric*

But perhaps the most interesting example of this continuity of thought within a developing type form comes from a scheme that we have prepared for the Royal Scottish Academy of Music and Drama in Glasgow. The site is a rectangular area of ground within the grid plan of the city. One frontage is on a main street: another faces a street that already accommodates the rebuilt theatre used by Scottish Opera which will, of course, have its links with the Academy of Music and Drama. A third side of the building will be bounded by a service road.

What is however important in relation to the design of the building is that the site is large enough to allow a spreading form of building in which load-bearing brickwork can again be used to secure good sound insulation in all the critical areas. The programme indicates that a very large number of rooms (particularly all the practice and teaching rooms) require this and again it has been decided that the maximum number of such rooms should have natural light.

The programme sets out the accommodation for two main teaching schools, Music and Drama and some special provision for opera. This gives rise to three main areas for auditoria: a theatre connected to the School of Drama and Concert Hall for Music, a laboratory theatre area and a similar area for opera rehearsal. Supporting this accommodation are all the teaching and backstage areas for drama which include a TV studio, dressing rooms, and scenery and property areas with easy access from the service road. The School of Music requires a very large number of teaching and practice rooms, and rooms for ensemble, brass and organ. In all of these the problem of good sound insulation is critical. In addition to these teaching areas the Academy requires accommodation for administration, the service of meals and common-room areas of various kinds.

The building form has evolved from a series of studies. The main rectangular building block contains all the teaching facilities. A second and much smaller block separates out the administration, common and re-creation room areas. Between these buildings and connecting them, is the main entrance hall with its interconnecting staircase and lift. A shared lecture

168

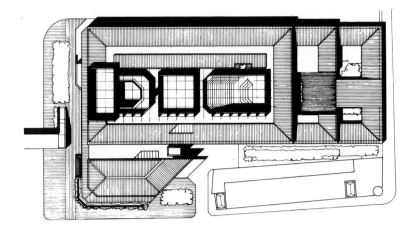

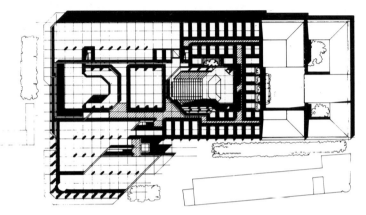

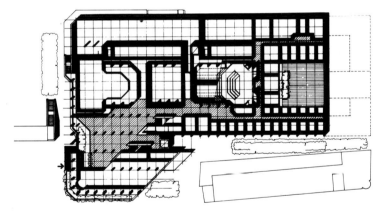

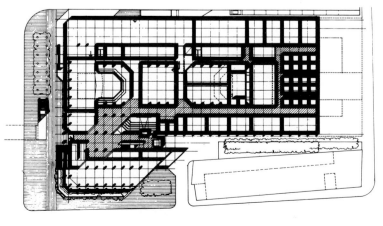

room and the main lavatory accommodation is accessible from this common area. The whole of the accommodation is planned within a block three storeys high, which in certain areas is reduced to two and in one practice room area one single storey only. A square grid as used in the Cambridge School brings all the room sizes into some kind of order and allows the specialized constructional system to be effective.

The planning is simple. Within the three-storey teaching block, all the main auditoria and the larger teaching rooms form a spine running along the centre of the plan. All these rooms are large and have a special volume suited to their purpose. The volume of the theatre and its stage rises through the height of the three-storey building and projects above the general roof level. So do the superimposed volumes of the laboratory theatre and over this the opera rehearsal room. The concert hall is entered at first-floor level and again its volume forms the third element in this spinal chain.

The general planning has to be thought of also in section. The principal accommodation for the Drama School is generally at ground-floor level. The accommodation for music is over this, at one end of the plan. All students enter the building at ground-floor level and move into the common entrance hall which connects the departments and the administration block. The public enters separately at first-floor level and has access directly to a common foyer which connects the public entrances to the main auditoria. The library is placed over this foyer.

The planning of the teaching block in fact forms three main and parallel bands of accommodation. First there is the frontal or access band, second the central spine of auditoria and third the servicing band which contains the dressing rooms and behind this the scenery, workshop and television studio with direct access from the service road. This gives a very convenient room to room arrangement within departments. But it also provides great flexibility of use: the dressing rooms for example, although they are associated with the two Schools, form a continuous band of rooms readily accessible from the theatre, concert hall, laboratory theatre or opera rehearsal room. Within this general layout each area of the plan can have its appropriate height.

The architectural form arises from, and in turn gives shape to, these considerations. It reflects the main content of the plan with its large-span volumes of the central spine and its load-bearing, sound-insulating construction elsewhere. The internal volumes suggested by the roof structure and the structural nature of the surrounding rooms give this building its special character. The load-bearing cross wall system, which is so prominent an element in the planning of the supporting rooms, projects outwards into the buttress or pier elements, which again have their sound-isolating function. This system runs right round the building. On the main frontage it is developed into a portico and gives a special emphasis to the point of entry. The external flight of steps leading to the public entrance and foyers forms the base to this pier and portico system. It establishes the division of uses within the building itself and a special scale in relation to the architecture of the street. This special scale, which is a feature of so many of the older buildings in Glasgow, is given an added point by the set back of the frontage itself and the broad pavement which this creates. This setting will have special paving and will be enhanced by the planting of trees. Again, as at Cambridge, the whole of the exterior facing is in buff brick which will harmonize with the colour of materials used in adjacent buildings.

Royal Scottish Academy of Music and Drama
Opposite page *model*
Left *plans*

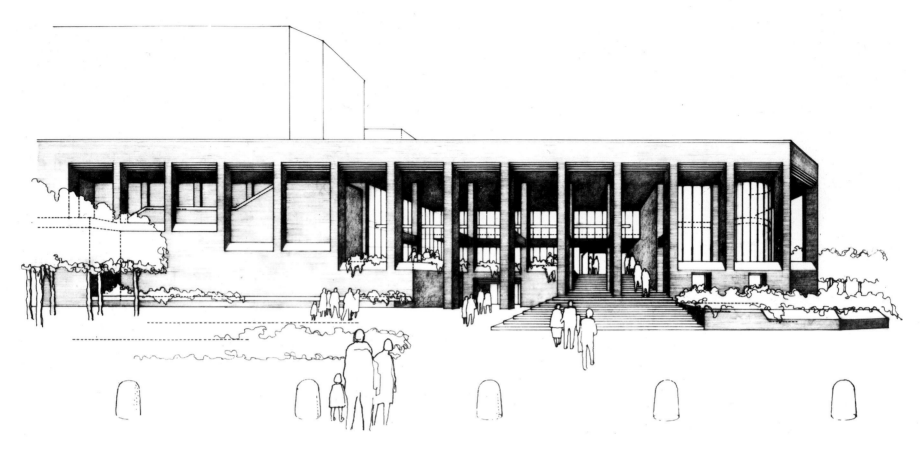

Royal Scottish Academy of Music and Drama
Above *perspective of main entrance*
Below *axonometric of auditoria*
Opposite page *model*

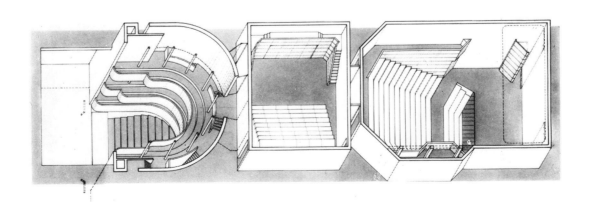

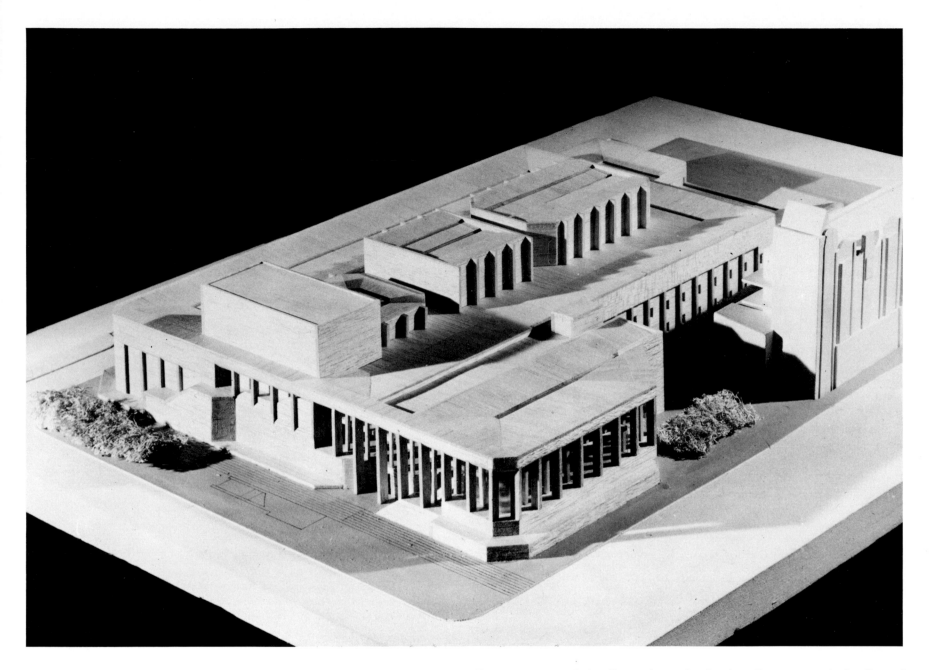

It seems to me to be quite clear that this building is an elaboration and development of a line of thought, within the study of a particular type of building, that is demonstrated in a number of designs produced in the studio over a considerable number of years. It suggests at least a continuity. The structural systems that have been accepted are simple. From the point of view of technology some things, for example the general acceptance of air-conditioning in interiors, have been deliberately avoided. That is not to say that they are not appropriate elsewhere or that I do not admire the skilful use of an advanced technology. These technical differences represent the separate streams of thought that I mentioned in my introductory note and are in my view the evidence of the broad front on which a developing architecture is being established.

From the point of view of my own studio, we have simply worked with the technical methods that seemed appropriate and we have avoided anything which appeared to complicate a direct solution to the problem. Every architect has his own particular approach to the design of buildings.

These notes and the illustrations simply describe one particular line of thought that we have followed through, I hope consistently, within the limits that we have accepted. What this certainly has produced is a developing and flexible system that we know and understand: indeed we have sometimes called it our vocabulary and, if we are building up a developing language of architecture, perhaps that is important.

NOTES

1   Some of the ideas in these notes are described and elaborated in a book *Buildings and Ideas 1933–1983 from the Studio of Leslie Martin and Associates* Cambridge University Press, October 1983. My associates on the first stage of the Cambridge Music School were Colen Lumley and Ivor Richards. During the final stages of this building and in developing the scheme for the Royal Scottish Academy of Music and Drama in Glasgow, my associate has been Ivor Richards.

2   *Marcel Breuer: Architecture and Material. An Essay in Circle* (p. 194) published by Faber and Faber, 1937, reprinted 1971.

3   Leslie Martin, 'Notes on a Developing Architecture' *The Architectural Review*, July 1978.

Alison and Peter SMITHSON

THIRTY YEARS OF THOUGHTS ON THE HOUSE AND HOUSING

Fitzwilliam Lodges

Hunstanton Head Teacher's House

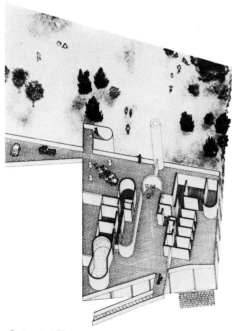

Cathedral Flats

INTRODUCTION

To remember the nineteen forties one has to invert many of the commonplaces of today . . .

the enormous numbers of cars and vehicles of all kinds constantly in movement and the noise level that is consequent – even in the country (1940s, few vehicles);

the number of domestic machines – appliances – and their operating noise (1940s, almost no domestic machines);

the quantity of personal belongings – even of the smallest children (1940s, few possessions);

the increase in bathing and the expectation of privacy and sole rights over toilet facilities (1940s, shared WCs, etc.);

the increase in bad public order, dirtiness and bad smells outside the house (1940s, the domestic streets so quiet as to be safe for children's play);

the expectation of a high level of heating and of its control (1940s, an open fire, generally only in one room);

and finally, the overwhelming acceptance of 'conservation' of good houses and an appreciation of housing built in the past (1940s, a general wish to condemn the past and to demolish old property and to make a fresh start).

A CRITICAL SURVEY OF OUR ATTEMPTS AT THE APPROPRIATE HOUSE AND HOUSING

THE NINETEEN FIFTIES

In the nineteen fifties we rethought our inheritance of housing forms, foreseeing that American advertising heralded the acquisition of domestic appliances. Most of our proposals for housing after 1954 attempted to hold to the tenets of this rethinking.

Fitzwilliam Lodges, Cambridge, 1949
Our first design of a steel and glass building, in the aesthetic of Mies van der Rohe which relates the organization of the parts of a building to its structure, the lodges are small essays acting in the general aesthetic.

Secondary Modern School, Flat and House, Hunstanton, 1950
The Caretaker's flat is built into the volume of the school whereas the proposal for the Head Teacher's house is a detached miniature of the school's architectural arrangement.

Cathedral Flats, Coventry, 1951
The ancillary accommodation is an exercise in coloured, rendered, non-structural walls, tailored to fit each room requirement in the aesthetic of the *Maison Savoie* of Le Corbusier; since both flats are free-standing they have private yards and terraces and enjoy all aspects and views. (This free-standing quality of the individual rooms was to lie dormant in our aesthetic until 1964.)

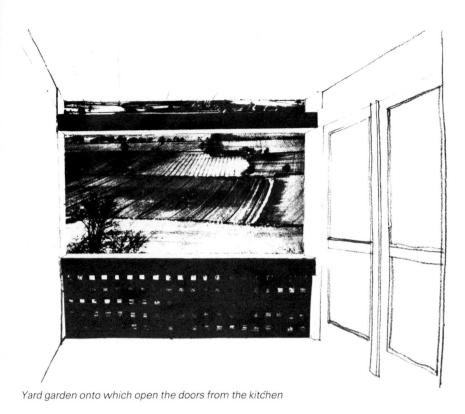

Yard garden onto which open the doors from the kitchen

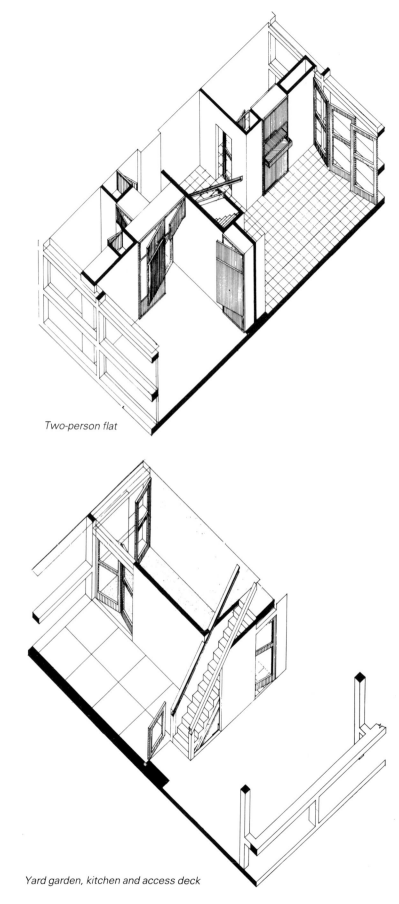

Two-person flat

Golden Lane, Deck Housing, London, 1952

1952 exploded with a full-blown idea about how dwellings should be organized; both in themselves and in relation to the group as inner-city housing. The aesthetic is now fully our own – *New Brutalist,* but before the term was invented. With this proposition the term 'deck' entered into the architect's (and commentator's) vocabulary: we believed all dwellings should have their access off some safe connective space where the first social contacts outside the home could take place. We also believed certain other criteria should be found in housing built at city densities: identity for the individual dwelling, for example, achieved not by pattern-making on the facade but by the reverse technique of quietening the architecture so that the scale of the whole could carry the tenants' choice of lamp-shades, curtains, etc. Offered is a robustness of detailing and a high degree of resistance to dirt and proletarian use – a necessity observed in Lubetkin's Busaco Street Housing. Golden Lane responds to an awareness of the apparently contradictory factors regarding people's belongings . . . that the household might want bicycles, pigeons, plants 'up in the air' beside them in high-density development; that is, the sense of disassociation which high density might induce requires more, not less, servant spaces; that outdoor extension is necessary to the proper functioning of the home; hence provision of yard-gardens . . . yet the built-in fittings envisage a throw-away attitude to clothes and so on: an attitude notably consistent with the spirit of the Modern Movement in architecture (as against a Victorian-style hoarding of goods). The freedom offered is the dwelling's response to basic needs . . . privacy for the individuals that make up the family, hence the concept of the 'parents' unit' (living, kitchen, parents' bedroom and bathroom) and the 'children's unit'.

Thus is offered choice and so a possibility for association within a piece of city . . . this is the use of high density to widen the experience of life on offer in the big city; in fact, it is only with high density that certain sorts of choices can be made available.

Yard garden, kitchen and access deck

Colville Place, a Terrace House, in a Pedestrian Passage, Soho, 1952

With this house, intended for the architects' own use, the *New Brutalist aesthetic* is announced. We are here concerned with a rethinking of the meaning of rooms as well as with their fresh arrangement in response to particular needs on a tight city site; we come up with: a poem of a stair; an attitude to equipment – for example, presenting equipment on walls on individual fixings, without 'designed' casings, if necessary exposing the basic mechanism so as to integrate the object into the general aesthetic. This is the realization that the changing appliances supply all the interest a house needs or that small spaces can support; the spaces respond to a common-sense grouping of functions; thus, one of the largest spaces in the house is allocated for storage of objects in secondary or interim use, a room with much the same meaning as a library.

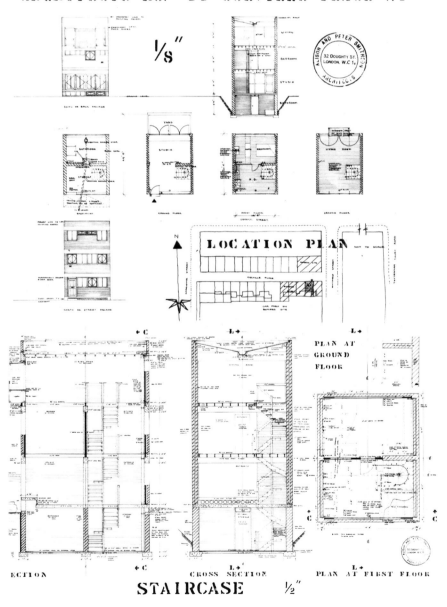

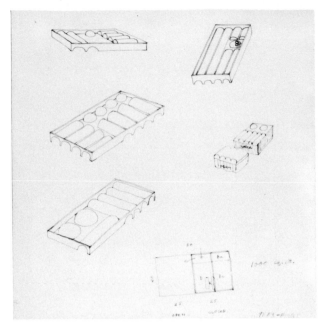

Tea-Tray Houses, sketches

1,000 Square Foot House (The Tea-Tray House), Highgate, 1952–53

Basic accommodation for the new couple . . . a second attempt to build for ourselves. The modulation of the 'flat' roof gave identity to the rooms covered so that the roof again becomes some form of 'hat'. This roof-as-hat idea ties in with other studies in the same period which investigated the meaning of subdivisions, while also avoiding taking the right angle for granted (about this time Le Corbusier's *Poem of the Right Angle* appeared).

Dwellings in a Hospital Compound, Doha, 1953

A group of houses for inhabiting a hostile environment, designed so that the result falls neither into sophistication born of ignorant insensitivity to location nor perpetrates, through over-enthusiasm for a peasant-past, what later became known as *casbahism*. Seven foot high compound walls are the first climate barrier; 'materials as found' are used for the building's mass to stabilize the internal temperature of the dwelling.

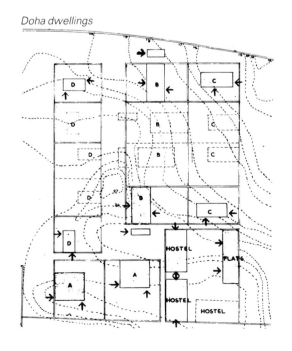

Doha dwellings

House Extension, Alderbury, 1953

The stepping-up in scale of an existing detached cottage to receive furniture of a more generous way of life.

Bates House

Bates House, Burrows Lea Farm, Shere, 1953

The solid-pierced-with-voids house — offering that sense of protection psychologically necessary to some people — is the Bates House which tries to extend the functional tenet to the positioning of the 'voids' in the 'solid'. The way each room needs to be looked out of, needs to receive light; the way each view should be appreciated, allowing a sense of the domain to be 'possessed' by those moving about the house.

With the outside walls so ordered it is then obvious that the 'apertures' — fittings and cupboards — inside should also be brought into the discipline; these 'apertures' in the 'thick' internal walls went to the side where they were most valuable — one side a drawer, or desk, and the other a wardrobe or a dressing table. This left perhaps one otherwise plain wall in a room; here a bedhead niche or well-disposed built-in lighting carried the discipline round.

In addition this most ambitious house tried to achieve:

a house-to-land relationship appropriate to retired people (a yard under the cantilever of the house and the drive make the relationship at ground level; the car passes between outside and inside like a boat into a dock);

a house-to-view relationship achieved in the disposition of its parts (the connections being the 'eyes of the house': the windows);

immediate extensions to the dwelling spaces (yard and terrace at upper level) allow the landscape domain to be enjoyed (without it requiring to be gardened);

subdivision into parents' and grown-up children's domains as self-supporting areas within the house;

everything organized for a leisured, servantless way of life accomplished through appliances and efficient heating, ventilating, and plumbing (that is, an American standard of comfort, in 1953 more or less unknown in England).

Staff Flats, Kampala, 1953

This, our second attempt at the 'hot-climate-tempering' dwelling, establishes its private world behind the *brise soleil*, under the roof umbrella.

Caro House, London, 1954–60

By being left 'open' to growth and change at every stage of development, the incremental conversion of a stable into a house ... sculptor's studio and yard ... children's domain ... painter's studio.

The Chance Glass Flat, 1954 and Mammoth Terrace House Conversion, London, 1954

The programme, the manipulation of external light by glass to carry light deep into the interior, is extended by us into an acceptance of responsibility for renewal inside old houses. This is to be achieved by raising the standard of the basic equipment: such simple things as one's own bath instead of communal facilities to satisfy the aspirations of the income group concerned — single people working in London offices.

The regeneration 'cell' is a prefabricated, smooth-edged container of the bathroom; carrying all services plus the warm-air heating trunking and the extract system necessary to an internal bathroom and kitchenette.

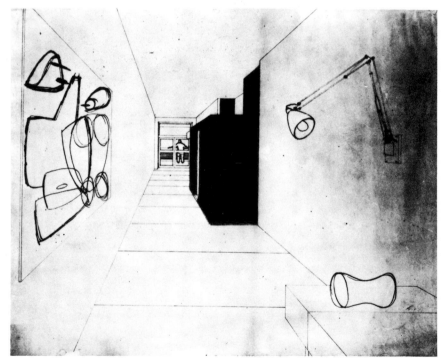

Mammoth Terrace House, interior

Limerston Street, London, 1954

Adjustment to new use by 'decoration' with the ephemerae of passing interests, the season, the travel acquisition (an exciting new option in the 1950s; what might be termed 'the retrieval aesthetic'; before the discovery of the world's remaining folk and before 'new brutalist' objects were made commercially available through Habitat ... the connection is direct, Conran was a student at the Central at the time P.S., Paolozzi, Henderson, Pasmore and so on, taught there).

The art of occupation-in-change, appropriate to our rethinking of housing and patterns of association.

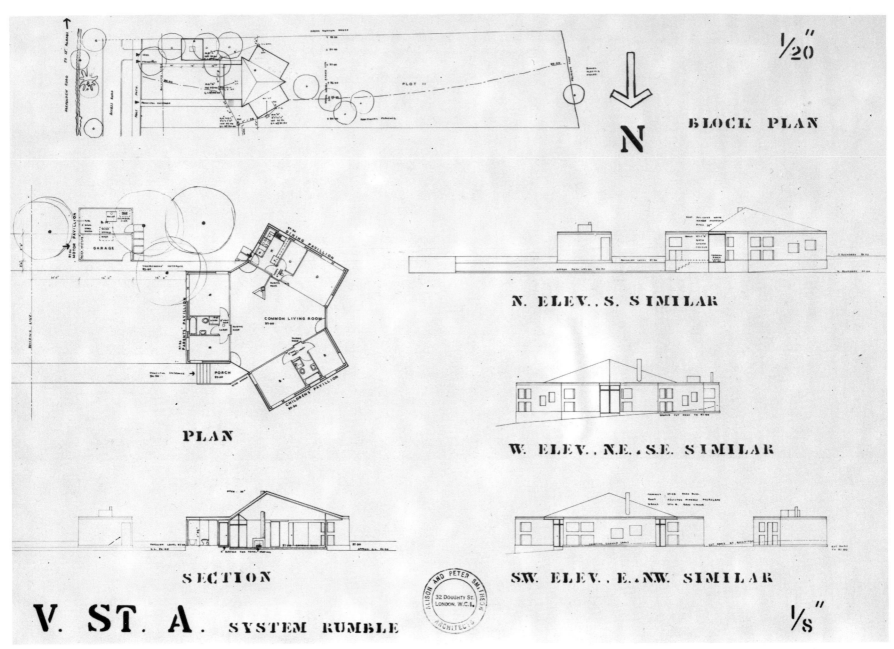

BLOCK PLAN

½₀"

N. ELEV., S. SIMILAR

W. ELEV., N.E., S.E. SIMILAR

SW. ELEV., E., NW. SIMILAR

PLAN

SECTION

V. ST. A. SYSTEM RUMBLE

⅛"

Rumble House, Home Counties, 1954
Use of the possibilities inherent in a sliding form-work system (and not stepping outside its discipline) to make a new grouping of rooms; a system-build house that could take up a variety of connective relationships.

Cheddar House of Cheese, 1955
Again the 1,000 square foot house permissible at this date; containing the basic cells, the basic spaces necessary for the young couple.

Cheddar House of Cheese

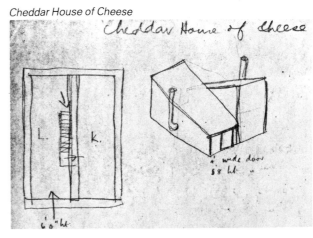

Team 10 Valley Section Housing: Studies for Hamlet, Village, Town, City: North of England and London, 1955–56

These studies were of housing groups appropriate to different sizes of community; each appliance-receptive dwelling so calm externally as to recede into its environment yet through its quality, be able to re-energize its context:

Galleon Cottages, for the Dales' hamlet;

Fold Houses, for infil at the end of 'lonnen' in Dales' villages;

Close Houses, riding the landscape for the renewal of small, northern, industrial towns, still compact in their landscape and still strong in a community sense born of past travail;

Terraced crescent housing, for infil in London; faced as a 'scoop' to the sun; thus able to give directional, identifying sense to a scatter of odd sites where their minimal shadow – due to the scoop form – could fall on roads or railways.

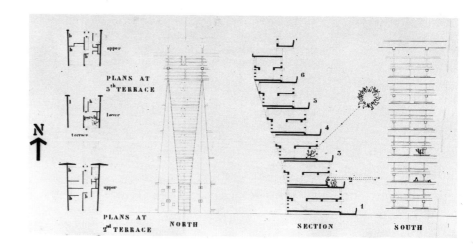

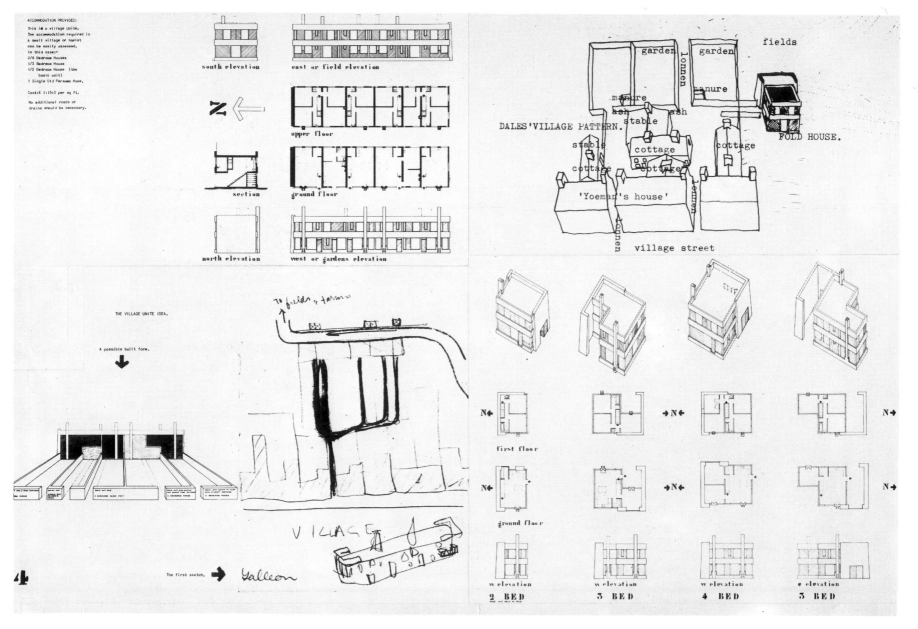

Sugden House, Watford, 1955–56

The first design for the Sugden House is related to the 1955 Close Houses, utilizing functionally placed identical windows: the best proportioned of the client's chosen manufacturer's standard range. By client choice, the final design – although it retains the parallel walls with simple timber spans between of the Close Houses, as well as the window positioning based on views and sun – has various combinations of standard windows and a conventionally ridged, double-pitch roof. The hope for use internally of a Canadian coloured wood preservative is represented by black-stained door frames (French polished).

House of the Future, Olympia, London, and Edinburgh, March and Spring 1956

Since our opportunities to build come so rarely, we always seize exhibition opportunities to project our ideas beyond our aesthetic – as if our ideas had already leavened the situation – and making correspondingly advanced assumptions for every field of endeavour likely to affect the house – we step into the Future. In 1956 we wrote:

'The rooms flow into one another freely like the compartments of a cave, and as in a cave, the skewed passage which joins the compartments, effectively maintains privacy. Each compartment is a different size, area, height; a totally differentiated shape to suit its purpose. It is the first exercise in the appliance way of life and tries to show the architectural consequences of the disintegration of the kitchen effected by packaging, food treatment, mobile appliances, etc. Furniture is designed with the same assumptions.'

The House of the Future indicates the prefabrication of a house whose inside and outside skins are such that walls, floor/base, ceiling/roof are thought of as one surface and joints are gaskets located in this continuous surface so as to function most efficiently, as well as modulate. Lighting, heating, and the air-conditioning are integrated in the prefabrication and the surface modulation.

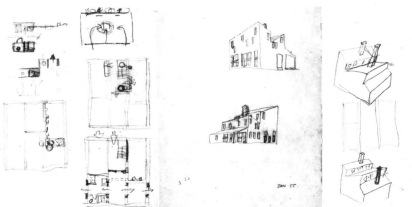

Sugden House

House of the Future, entrance door

This is Tomorrow, Patio and Pavilion

This is Tomorrow, Patio and Pavilion, Whitechapel Art Gallery, September 1956

A re-statement of our position as well as our attitude to objects and possessions – 'the retrieval aesthetic'. We were therefore concerned, in our separate disciplines (architects; sculptor, Eduardo Paolozzi; photographer, Nigel Henderson), with presenting our images and ideograms to satisfy the different aspects of the needs of man. To best achieve these ends we worked on a kind of symbolic 'habitat' in which can be found, in some form or other, the basic human needs – a piece of ground, a view of the sky, a sense of absolute privacy, the presence of nature (and of animals when we need them), as well as the symbols of the basic human urges – to order (and extend control), to stay (and to move), to make an art of inhabitation.

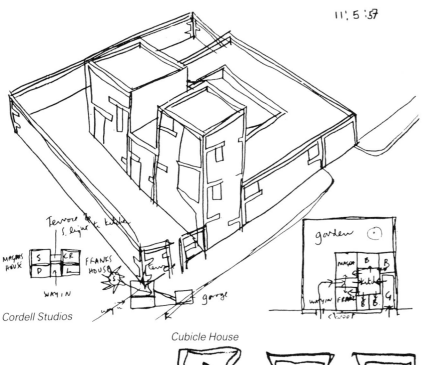

Cordell Studios

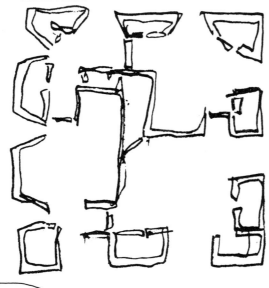

Cubicle House

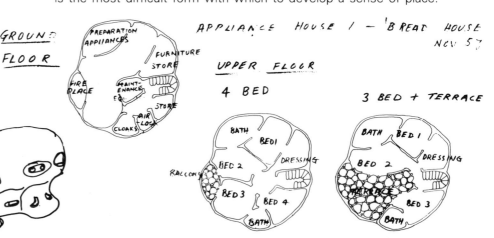

Appliance House 1, Bread House

Bark Place, Bayswater, London, 1956

External storage accommodating the car, the garden equipment and storage, heavy accessories for the house, etc., formed so as to extend the garden of a small house.

Cordell Studios, London, 1957

Ideogram only for fun, a gift drawing . . . a quadripartite dwelling with three work areas plus a 'house'.

St Peter's Villas, Hammersmith, London, 1957

Internal storage extension by cantilevering into the over-stair space of a London house.

Criteria for Mass Housing, 1956–57

The failure of Government and other authorities concerned with housing and planning to give either fresh direction or leadership in the matter of how houses should be grouped, or what amenities might be provided by such grouping, caused us to write down Criteria for Mass Housing. Any such criteria should be simple to meet, analogous to Byelaw and Public Health requirements (which seem to be adequately applied and met). Various attempts were made to establish Criteria for Mass Housing, for example at the *C.I.A.M.10* meeting at Dubrovnik it was thought the key might lie in how access was provided to buildings and in the way people moved about their larger group – the region, the city, the town; or simply in the hierarchy of amenities. But the problems raised by a child of one, starting walking and wanting to play in a street becoming visibly busier, returned us forcibly to those basic principles embodied in the Golden Lane proposition and the following of this train of thought produces the Criteria for Mass Housing.

Cubicle House, 1956–57

After the House of the Future and the Team 10 housing studies of 1955–56, new ways of getting at the archetypal house are sought; the series of Appliance House studies are the result. The Cubicle House is the start of an idea for an appliance house with the 'put-away' aesthetic.

Appliance House 1, Bread House, 1957

The appliance way of life suggests an entirely new sort of house: the working out of an idea in such plastic forms causes us to shed automatically all preconceptions. This is also a move towards a pattern of grouping for the favoured isolated house that allows the garden spaces to flow together uninterruptedly so as to be capable of genuinely being the supporting 'garden-like' matrix of the suburbs. Our attitude is not to reject the idea of the detached house because it commonly uses the most devalued symbols and is the most difficult form with which to develop a sense of place.

Snowball House, 1bis

The 'put-away' aesthetic involves a change of location of the storage spaces or appliance cubicles, so that the enclosing shells punctuate the house area, insulating each effectively from the others. Cubicles contain all connection points for services and mechanisms while the outside of the shell, in its folds, carries the lighting for the house-space proper as in the House of the Future.

Appliance House 2, Strip House

Appliance House 2, Strip House, Home Counties, 1957–58

Application of the foregoing ideas to possibilities of late 1950s technology and the introduction of mechanized building operations, at their most rational, brought the rectilinear, appliance, row house; the self-imposed programme is as follows:

capable of mass-production;

every detail in present-day technology;

the single dwelling to be capable of grouping in a series of numbers;

that special account should be taken to integrate electrical runs (to advance out of the 'wires-age'), etc., that is, complete services integration;

to imply a range of sorts of garden/outdoor extension to the dwelling;

to have a high degree of privacy;

to make the connections to other routes – to the town or region – for the sole use of the group, indicating by their form, a bid for safety, quiet and cleanliness;

to present a certain glamour;

to require no maintenance (keep fresh-looking with just cleaning down, like a car);

to be highly insulated against cold weather.

The desirable pattern of use we see as:

covered access from car;

access to all rooms without passing through others;

to look out on to something decent;

children's space implied in the group;

adolescent space implied in any larger grouping.

The general aim is to regain as much as possible of the house as usable space by continuing to assume that appliances do away with the need for the work space in the old sense.

Appliance House, Retirement House in Kent

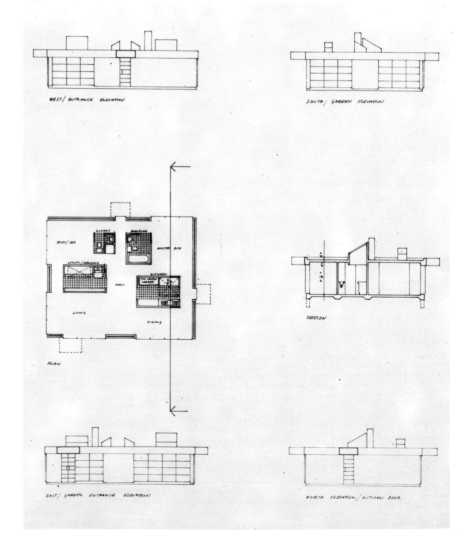

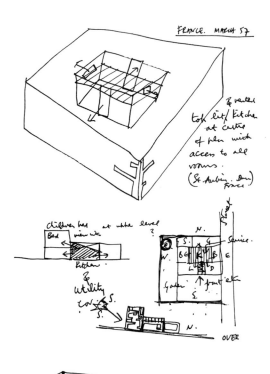

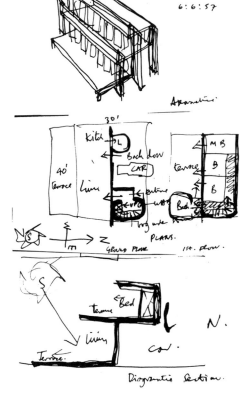

House Idea, 1957
The house for the enjoyment of the seasons, the climate.

Climate Houses, Paisley and Hampstead, 1957–60
The climate house subjects the British house to a climate analysis.

Boodles Club Residential Rooms, 1959–64
A classic 'appliance cubicle' plan arrangement with the servicing of the cubicles from the access corridor.

Paolozzi House and Small Retirement House, Kent, 1959–60
Two attempts to build the appliance-responsive house inexpensively.

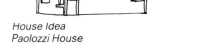

House Idea
Paolozzi House

Climate House

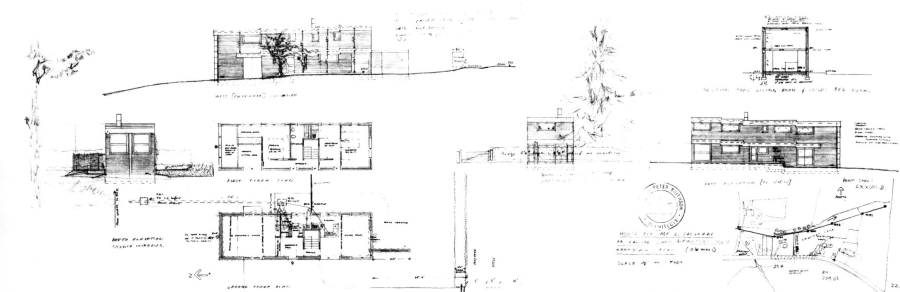

Wayland Young Pavilion

Churchill College, ground floor lawns and protective belts of hedges

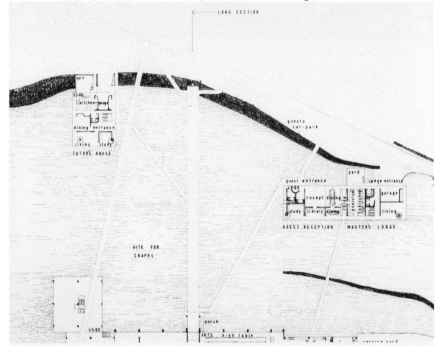

Wayland Young Pavilion, Bayswater, London, 1959
Two appliance cubicles and a single mature tree give this small garden pavilion its plan ... an arrangement that has successfully adapted itself to three generations of use.

Sets and Lodges, Churchill College, Cambridge, 1959
Churchill College is intended to serve the students and Fellows of a new enlightenment. The grouped college sets and soft-form appliance rooms (the closer to the user, the softer the form) are a modified version of the first Cubicle House. The skin, on parallel tracks, offers options of sliding windows, controllable ventilation, sliding shutters ... much in the same manner as does an Indian railway carriage.

In the two lodges, added to this general idea, is the idea of a special freedom of the outside skin, such as is found in the early English Renaissance house plans of Roger Pratt, so that the skin appears free, continuous, and wrap-around, suggesting academic perambulation when the communication doors are open, and offering the writer or reader a wider view of the well-kept grounds: the occupier is made constantly aware of the garden as an encircling pleasure. This expanded view outward, the extension of a particularly English enjoyment of the garden, is to satisfy the newly re-found sensibilities as to the nature of places.

Gordon Place, London, 1959, Provost Road, London, 1960
It should be possible to bring to bear on conversions of existing houses sufficient of one's discipline to recreate spaces appropriate for the occupier's period of inhabitation ... Charles Rennie Mackintosh managed this.

Upper Windows, Koestler House, Alpbach, Austria, 1960
All-weather French windows within the gable, up-dating the traditional climate-responsive facade.

IN RETROSPECT
Looking back at the nineteen fifties, it is clear that we worked like any young architect, excavating our immediate inheritance – Mies van der Rohe, Le Corbusier, Lubetkin – and for us in this period, excavation included our industrial inheritance, at that time being disparaged and destroyed. But we try to learn these lessons from our past without any failure of nerve on our part; we believe it is not only possible to equal the invention of the past with appropriate inventions for our own period ... it is also our duty not to live off built capital.

THE NINETEEN SIXTIES

With the Churchill College design, made in 1959, and that to be made for Upper Lawn, 1961–62, our focus of attention begins to shift to the 'skin' – as protective membrane and as indicator of the bounds of what is private and what is public, with all the degrees of privacy in between.

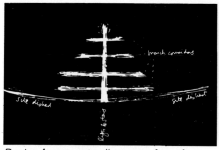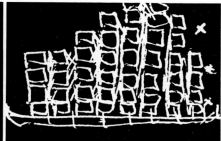

Duplex Apartments, diagrams of tree form

Duplex Apartments, Hove, 1960

This fifty-apartment tree for Hove grew from the wish not to take away from any house in the hinterland the sense of the sea. The circulation was like the timber of a tree; the lifts and stairs as the trunk, the central corridor the branches, getting shorter the higher they occurred every third floor. (This form also economically satisfied the Means of Escape.) The foliage wrapping the branch–corridors was duplex apartments.

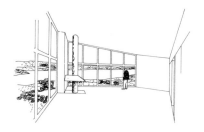

Losey House

The Losey House, Wales, 1960

Nestled between the sea and Snowdon, enabling the occupants to appreciate all the variety of the encircling views, a house whose wings, for parents, for children, small family or large gathering, seemed to grow out of and return into the existing stone field-walls.

Cliff House, Wonwell Beach, Cornwall, 1960

On a shelf above the shore, at the mouth of an estuary, replacing a derelict fisherman's house; on a family estate the planners now sought to control; a holiday home to be opened and closed easily; seen as being a self-protective form akin to *Peggoty's House*.

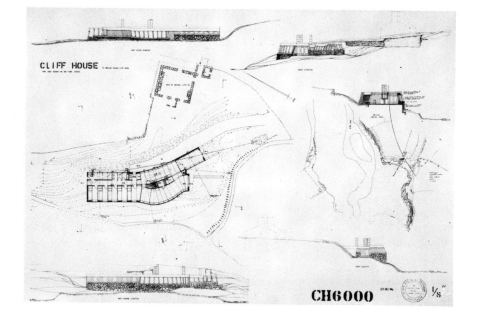

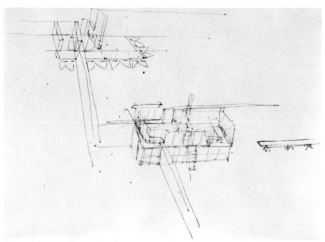

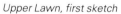
Upper Lawn, first sketch

Upper Lawn, photographed in 1963

Upper Lawn, Fonthill, 1961–82

A thin-skinned structure in a compound, openable-up at garden level . . . like a 'scena' changeable for the acts of inhabitation at the different seasons of the year. First use of the square on the diagonal in our work is in the ground-floor reinforced concrete columns with rounded corners, for they are close to the touch. This is a true solar pavilion, designed to capture the sun from its rise to its setting, especially in mid-winter . . . a theme which re-appeared at the end of the 1970s.

Priory Walk, London, 1961–71

Conversion of the top, attic level of a tall terrace house to a single space as a family kitchen-dining room: to be in the sun and enjoy all the back gardens laid out as a collective mosaic.

Robin Hood Lane

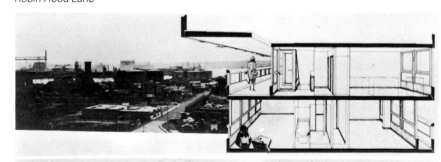

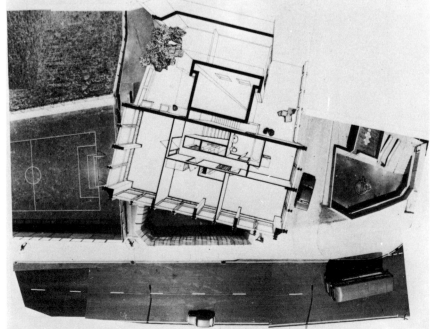

Manisty Street, 1962–64,
Robin Hood Lane, 1964–72,
Tower Hamlets, London

An application of the ideas developed in the Golden Lane deck housing studies in 1952–53; but of course the situation, particularly with regard to vehicle noise from the surrounding road network and amount of private car ownership had changed dramatically in the intervening twenty-year period. The site configuration of the buildings and the modelling of their skin to respond to the new circumstances are the beginning of an aesthetic of layering.

British Embassy Residence and Lodges, Brasilia, 1964–68

A protective double roof and continuous overhang as a protective skin offered equable internal conditions for living on the high plateau of Brazil. Long spans allowed a free grouping of rooms; as if the rooms were decorative objects arranged on shelves. The majority of rooms opened on to terraces that allowed expansive views, over the secure garden, to the lake.

Housing, Street, Somerset, 1965–67

Area studies of mixed layouts of Close Houses and sophisticated Galleon Cottages where the car entered for off-loading direct into the house, an attempt to fit the town-house requirements of managerial staff. On sites lacking any view out, patio houses are developed from the House of the Future.

Burleigh Lane Housing, Street, Somerset, 1966

A cluster of detached houses; with car-courts on the roadside to the north and private terraces, protected from winds, on the south, garden side. The 'V' plan geometry of the houses, in repeating along the street, provides the necessary protection and enclosure as well as fresh grouping characteristics.

British Embassy, Brasilia

Patio Mat-Housing

Burleigh Lane Housing

Tutor's Flat, St Hilda's College, Oxford, 1967–70

A tutor's flat at one corner of the Garden Building is like a musical variation on the building's theme.

Patio Mat-Housing, Kuwait Old City, 1968–69

With accurate climate information available to us of such a rugged climate as Kuwait, we were able rationally to extend the language of the traditional high-density urban dwelling, offering patio mat-housing of two storeys and roof terraces, raised on low pilotis that gave undercover unpacking/entrance/car parking, whilst creating at ground level a breezeway in from the Gulf to the hinterland.

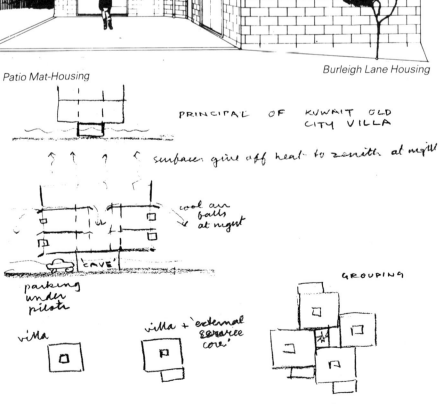

PRINCIPAL OF KUWAIT OLD CITY VILLA

surfaces give off heat to zenith at night

cool air balls at night

'CAVE'

parking under pilots

villa

villa + 'external terrace core'

GROUPING

THE NINETEEN SEVENTIES

In this decade the layering of the facade skins finds response in the plans and sections of the dwellings . . . these ideally have a varying density within their depth . . . open-up to sun and quiet . . . close-down against cold and noise.

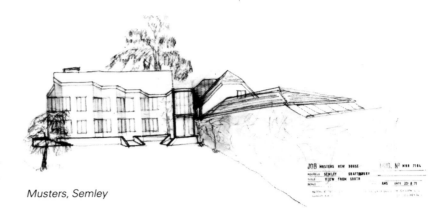

Musters, Semley

Musters, Semley, Wilts, 1971

House emerging from existing old barn into paddock; a house to accept the outlook into every room; a house for the English climate.

Gilston Road, London, 1971

A layered conversion: children's level, parents' level, studio levels; detached London house that and therefore allows two characters, two entrances, the house's domain and the studio's.

Cherry Garden Pier, Southwark, 1972–74

A very special site in that it requires the re-use of deep foundations of an existing warehouse from the late 1940s. The first project – a chevron plan where the windows were planar with the river views – is amended to a plan with a more cubic form. In this second version, the existing peripheral piles are used by 'cottages' around the site's edge and the interstitial piles by the studio/maisonette's 'flounce' around the base of the central block . . . any odd pile carrying an earth container for a tree. The adjoining sites have other architects; the shared hope being to make a compatible place within old dockland and to show, that with goodwill, architects can build places.

Cherry Garden Pier

Executive Flats, Battlebridge Basin, London, 1972–74

Over a building where a number of professional people could have their practices, some would also have their flats. All flats have a facade to the basin, the end flats also have a street facade. In all flats the rooms encircle internal patios and so have both the pleasures of natural ventilation and a private view of the sky.

Kreuzberg Infil, Berlin, 1975

Infil to gaps in the periphery of a Berlin city block; extending the traditional language of Berlin construction but shifting its meaning and emphasis to our period. The existing facades of the block derive from the Renaissance, the internal structure is timber framing whose characteristics are completely German; it is this hidden language that is taken up again to provide light-reflecting, many-layered facades, allowing, as the outer layer, the flower boxes that bring the seasons into the depopulated inner city.

Kreuzberg Infil, Berlin

The Yellow House, 1976
The Upper Lawn pavilion-up-against-a-wall idea is further elaborated with a basket-weave, braced, skin.

Town Houses, Landwehrkanal, Berlin, 1976
Study for infil in Berlin adjacent to the Kanal. To avoid the locked-in feel of the old Berlin courtyards, this infil holds the line of the street but never closes the possibilities of views out from the centre of the block.

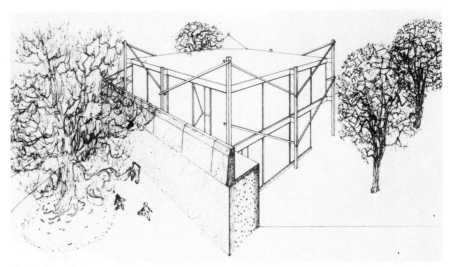

The Yellow House

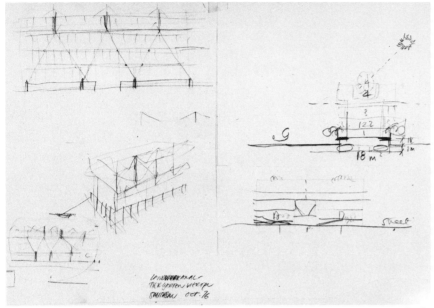

Town Houses, Landwehrkanal, Berlin

Llangenith, Wales, 1977
Additions to a village in a rural area; the clustered dwellings echo the building mass of the traditional farms; the interlocked dwellings offer a new social amalgam between local residents and incomers.

Worcester Moves West, 1977
This renewal study (made in collaboration with others) offered a family of urban-scale infil housing:

youth 'pads' for all young people wishing to live 'at the centre', mixing young typists, bank clerks, shop assistants with polytechnic students (the available hostel accommodation is all on the edge of Worcester);

starter flats of basic accommodation for the married young, accessed off covered ways that were extensions of the traditional covered ways of Worcester: ways where could be located launderettes, take-away food and record shops. The flats stacked 'back to back' in three walk-up layer-sizes (the smallest at the top) and by descending the slope to Dolday to the west, slip-stepping on their party walls, every terrace obtained views of the Severn and west or north-west and south sun;

old people's enclave, extending the urban language of the fine architectural quality of the historic alms houses of Worcester;

family town-houses, descending the western slopes to the Severn and overlooking the cricket ground; again each 'cluster' accessed off covered ways from their car access/garage level; the house plans cranked, responding to the slope, to allow all year south-west sun into the heart of each dwelling; the terraces and gardens descending to private outlets on to the pedestrian footpath provided a worthy neighbourly cluster to the Bishop's Palace adjoining the Cathedral standing above the slope southward;

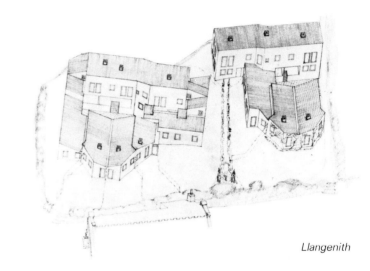

Llangenith

Worcester Moves West

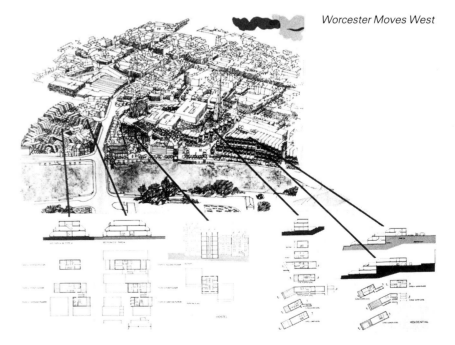

Millbank, Chinese axonometric

Millbank Housing, Westminster, 1977

Facing south across the Thames with a noisy road on the north side, this proposal is the start of the theme of varying the plan/section density from the 'soft' south skin inwards towards the 'hard' protective mass of the north skin.

House with Two Gantries, 1977

The town-house section explored as a scaffold for a layered display of the inhabitants' possessions/collection. The amount of space needed for storage of objects of an intermittent use which can change the way the house and its decorations are used seems to have never before been fully explored.

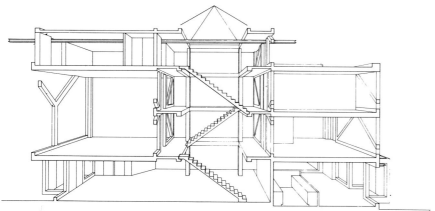

House with Two Gantries

Cookies' Nook, 1977

In the way that a *Volkswagen* can look appropriate in any city, this is the container-pad for any city. Intended for the peripatetic professional or academic to be able to put his hand on everything always in the same place; providing an adequate working environment, a home from home with appropriate appliance cubicles, within a prefabricated container-house for placing in cities throughout the world . . . only the view outside the windows would change and to appreciate better the best any city can offer, the stacking units can be slewed this way and that to fit on centre sites and work their way into the city and its pleasurable qualities.

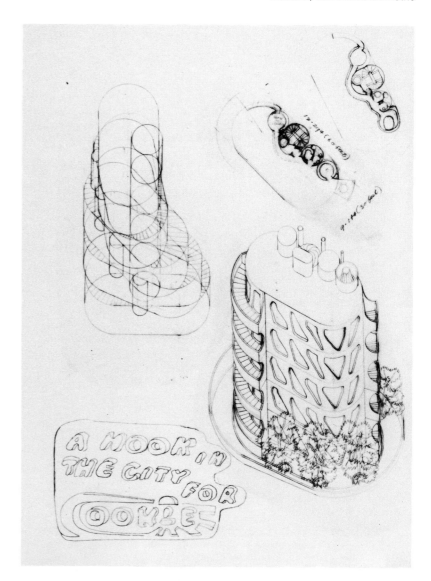

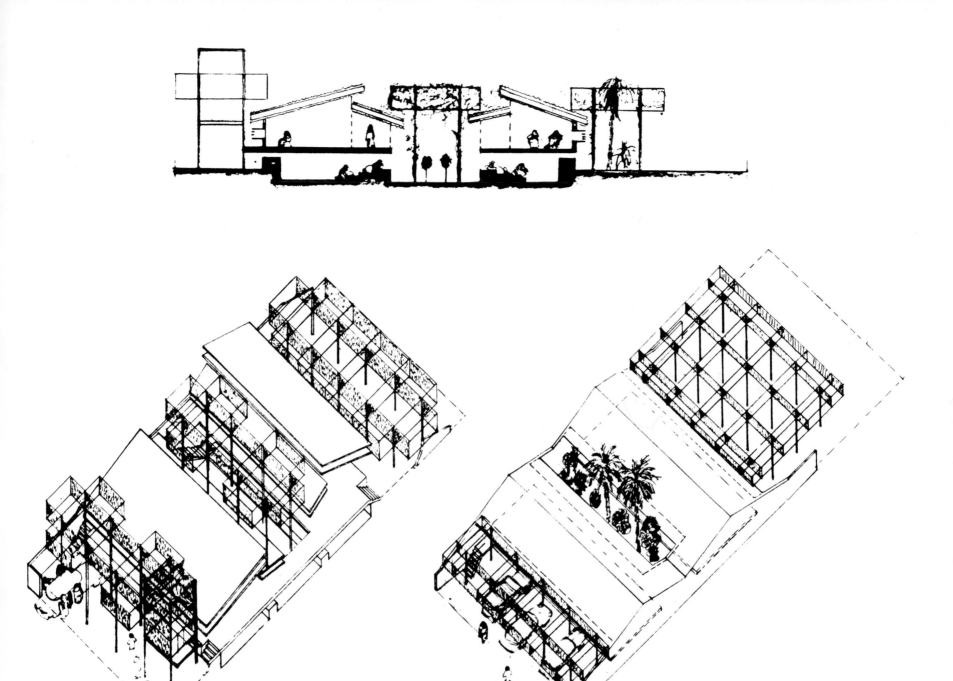

Forum's Cubitt Home

Forum's Cubitt Home, 1977

The idea of a man-made oasis created by a grid of high-level perforated screens carried on masts. The grid could take up many configurations, from the single double-stacked square, that makes a 'lattice' gazebo, through variations of over-courtyard screens that could carry climbing plants, to the grove that could shade a children's play area, or communal barbecue area. The four variations of 'climate' house explore the use of this shaded oasis idea.

Scholars' Residences, Pahlavi Library, 1977–78

Discretely occupying their own court and garden spaces at a corner of the complex.

THE NINETEEN EIGHTIES

In the nineteen eighties we may see the end of the period that Shadrach Woods identified as: 'ever more sophisticated machines inside ever more brutalized containers'; for although there is generally a cruder level of achievement in traditional on-site building trades, with much waste, there is much more sophisticated handling equipment and a wide range of building services components and control equipment of high technical standard, precise functioning, and long life.

With general availability of this control equipment we can turn again to the sun: in the Heroic Period of Modern Architecture the sun was an impulse towards the invention of the architectural language; in the houses and flats of the period, terraces, usable roofs, balconies and open-air rooms are all so clearly stated that we commonly regard them as purely stylistic devices but of course they are *machines* for receiving sunshine in.

At the time they were in tune with a general feeling that the sun did more than help to cure and help to prevent tuberculosis ... the feeling that to expose the body to the sun and to the elements is a good thing in itself.

Sixty years later that same impulse flows again, but starting from a different source ... for we all now want to catch, control and enjoy what sun we can: and concomitantly we all suddenly feel that pollution of all kinds is cosmic desecration and an inhumanity to man and nature.

Lützowstrasse, Berlin, 1980–81

The south face of this long protective housing along the Lützowstrasse has an 'energy-skin-space' whereby the 'rigid' cocoon of the dwelling is made capable of being burst open when summer comes; the dwelling shrinks back when winter returns. This summer extension (echoing the response to season of the old 'sommerwagen' of the pre-1914 Berlin tramway) as passive 'energy-skin-space' is something wholly new – an extra to the basic dwelling in terms of area. It is imagined that insulating blinds, with sun reflecting surface for summer, would be installed behind the external glass face; but as energy conservation technology advances, the market will offer to occupiers the possibility to add inside extra removable linings ... such as panels, or louvres, filled with glaucous crystals which absorb heat during the day (perhaps when the occupiers are at work) and are turned to give out heat into the dwelling at night. When such panels are closed tight they also offer their insulated face to the glass. The floor and side walls of the 'energy-skin-space' are reflectively tiled to bounce light into the dwelling in winter. In summer time the inner line of French windows folds away, greatly enlarging the occupiable space.

Most of the bedrooms are on the quieter, north, side towards the Landwehrkanal (and the Tiergarten) where the air is somewhat freer from pollution. On this thick north facade, shutters cover all windows at night. The reveals of all north-facing windows are a mirror-finish to catch east and west light and reflect it into the dwelling; for we believe a sense of the weather to be vital in north-facing rooms.

This housing along the Lützowstrasse is a very radical proposition, realizing a long-worked-for-sun-response format; with a plan of variable density; and also with a 'natural-seeming' level-to-level ascent.

To realize the innovative force of the idea of 'varying density plans and sections', of 'opening up and closing down' with ease, of ascent under full surveillance ... a range of formal inventions is needed and this is our objective for the nineteen eighties.

The Winter Garden

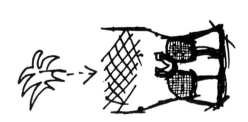

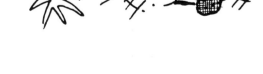

The Winter Garden, the Snug and the Cubicle Apartment, September 1982

To see to the horizon and down to the ground is a special pleasure in the city, but high-up the sense of exposure is sometimes intolerable so there should be somewhere inside to retreat to ... as into a shady corner of the garden as one would in a house on the ground. The winter-garden to see out from and the snug to retreat to are a high-up man's house's main elements ... for each the furniture needs to be quite different: sun and water-proof, lace-like, light filtering in the winter-garden; solid, divan-protection in the snug; the cubicle of racked storage and the cubicle vapour proof containing appliances, are the necessary life-supports.

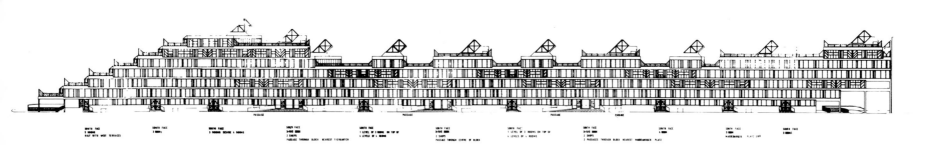

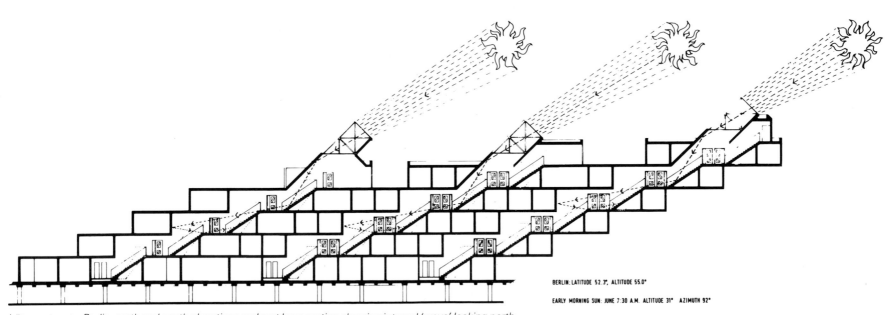

BERLIN: LATITUDE 52.3°, ALTITUDE 55.0°

EARLY MORNING SUN: JUNE 7:30 A.M. ALTITUDE 31° AZIMUTH 92°

Lützowstrasse, Berlin, north and south elevations and part long-section showing internal 'ways' looking north

James STIRLING

As a practising architect I want to talk about projects in the office, although I will briefly refer back to some of our earlier work.

On this occasion[1] I would like to consider our work in terms of the context, no context or anti-context, and relating this to earlier projects which I would categorize as either 'abstract' or 'representational'. 'Abstract' being design solely to do with the modern movement and the vocabulary of forms derived from Cubism, Constructivism, De Stijl and all isms of the new architecture. 'Representational' being design related to tradition, vernacular, history, recognition of the familiar and the more timeless concerns of the architectural heritage. I would claim that both aspects (extremes?) have existed in our work since the beginning, with projects being to one side, often to the exclusion of the other; more recently I think both aspects are being contained and counterbalanced in the same design.

NOLLI PLAN, 1977

In our piece for Roma Interrotta[2] – the architectural game of 1977, wherein the then superstars each re-did a piece of Rome as an update of Nolli's 1748 city plan – we made a veritable thesis of contextualism using buildings and projects designed in the previous twenty-five years,[3] integrated or juxtaposed, into and all over the Trastevere site.

To quote from the text which accompanied this proposal:

'This contextual-associational way of planning is somewhat akin to the historic process (albeit timeless) whereby the creation of built form is directly influenced by the visual setting and is a confirmation of, and a complement to, that which exists.'

Some examples:
A The waterside siting of our Oxford and St Andrews dormitories here sited overlooking the Tiber
B Olivetti's extension to a country house, here extending the Villa Farnesina
C Village project, here sited in the Roman campagna
D The Dorman Long and Selwyn College projects, both wall buildings, here reinforcing the alignment of the Gianicolo and Aurelian walls
E An interchange of monuments, Garibaldi on a horse replaced by JS on a birthday cake (fiftieth)

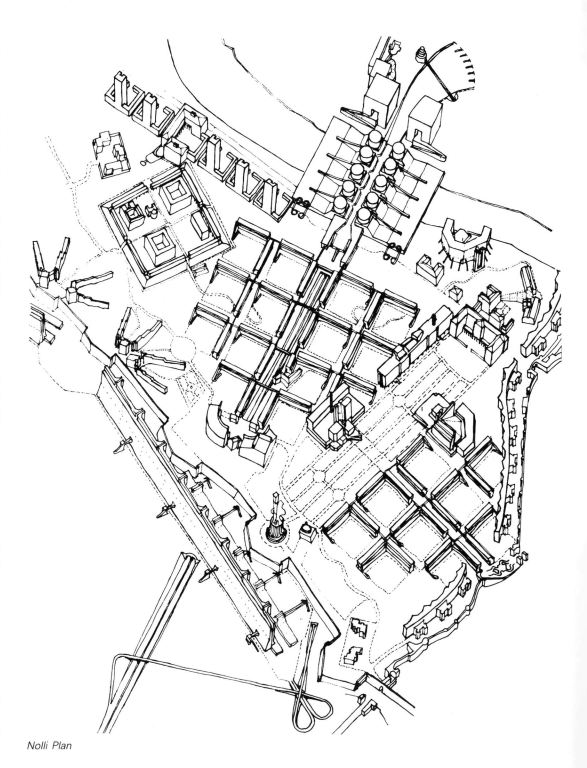

Nolli Plan

F The inclusion of a university precinct with our projects from Sheffield, Leicester, Cambridge, etc.
G Social housing in the form of squares and terraces (Runcorn) approximately equivalent to the residential mass that has' grown in this area of Rome since 1748

and many more, all with parallel references.

There was a larger objective, perhaps over-ambitious, to see if, by using our projects only, it was possible to establish the fragments of a city; to see if our collected works in any theoretical way added up to a tentative urbanism.

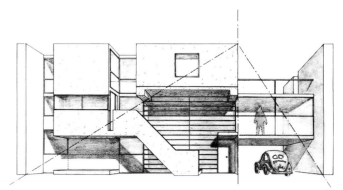

Core and Crosswall House

However, to go back to the beginning. In 1950 I finished at architecture school. Projects which followed included:

CORE AND CROSSWALL HOUSE, 1951

This scheme from the early 'fifties was for a single-family house and was unashamedly and totally 'abstract' – there was no reference to anything earlier than the twentieth century; in those days one believed that modern architecture could do it all.

Village Housing

VILLAGE HOUSING, 1955

Whereas this project of only a few years later was a theoretical study in how to extend the typical English village. It was 'representational', that is traditional, vernacular, and not dissimilar in appearance to High Street villages as they have existed since medieval times. There is perhaps more system although the building forms and their clustering are quite traditional, even folksy.

Mavrolean House

Leicester Engineering Building

MAVROLEAN HOUSE, 1957

Similarly, this project for a rich man's house in South Kensington was to have been built in an area of Edwardian mansions and was influenced by the form and materials of that context. It was 'representational' with veneered stone walls and copper roofs, internally it had traditional arrangements of conventional rooms.

LEICESTER, 1959

However, the buildings which were built (here the Engineering Building at Leicester University) tended to be in the 'abstract' category and where the surrounding context had little influence – maybe, here, because the site was in the backyard of the University, although we did build a tower with a view over a city park; perhaps in so doing turning our back on the University. I guess we judged the view of the park more interesting than that of the campus. The forms of the building were influenced by the 'abstract' vocabulary of the modern movement and it has been referred to as Russian Constructivist, Frank Lloyd Wright modern, etc.

WISSENSCHAFTSZENTRUM, BERLIN, 1979

The next but one German competition we also won,[8] for the 1986 Interbau where, instead of concentrating the new buildings in a single area as previously, the intention is that they will be sited in different parts of Berlin, where each can be beneficial, even remedial, to an immediate neighbourhood. This building for the Bonn Government is really a Think Tank but called a Science Centre — an Institute for deep thinking on matters of ecology, environment, sociology, management, etc. We have to reuse the huge old Beaux-Arts building (by the architect who built the Reichstag) which somehow survived the war. Having destroyed so much with post-war reconstruction, they now want to hang on to everything which remains. The site on the Landkanal is between the Fahrenkamp Shell House and Mies van der Rohe's Museum, and in this place of devastation we hope to create a microcosm — 'the fragments of a City'.

The Old Beaux-Arts Building

200

Reading from our competition report:

A 'The primary need is for a great multitude of small offices and a particular concern was how to find an architectural and environmental solution from a programme composed almost entirely of repetitive offices. The typical office plan usually results in boring box-like buildings and the banality of these rationally produced office blocks may be the largest single factor contributing to the visual destruction of our cities in the post-war period.'

So, I made an early decision that, whatever, we would break away from the office block stereotype and I said to those working on it: 'Make a cluster of buildings – take for instance a long bar, a cruciform, a half circle and a square, and juggle them together with the old building.' Quoting from the report:

'Our proposal is to use the three Institutes of the Centre (Management, Social and Environment) plus the element for future expansion to create a grouping of four buildings, all of which are similar, but different, and the architectural forms may evoke familiar building types, with each Institute having its own identifying building.

B 'Each Institute has two directors with complementary staff and a binary organization seems fundamental. As the buildings have symmetrical planning, allocation of rooms should adapt to this dual organization rather well.

C 'The new buildings cluster round an informal garden and the loggias and arcades which overlook this garden also relate to the cafeteria, the conference facilities and the old building. There is a free-standing Library Tower with a reading room at garden level.

D 'We hope to make a friendly unbureaucratic place – the opposite of an institutional environment, even accepting that the functional programme is for a great repetition of small offices in a single complex. Actually the whole can also function as a single complex as each building is joined at every level.'

The irony is that, while we have greatly varied the building form, we have maintained a characteristic of repetition in the wallpaper-like application of windows. A decision regarding the typical office room was that each would have a single centrally positioned window with flanking wall surface allowing for curtains, bookshelving, etc.

Externally, walls are rendered in *Putz* (stucco) with alternating colour, pink and grey, per floor. This project has now reached working drawing phase, but its start date has been set back into the following year's budget. We still hope to get it finished by 1986.

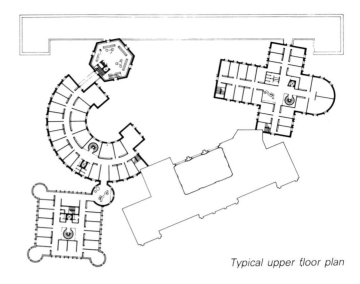

Typical upper floor plan

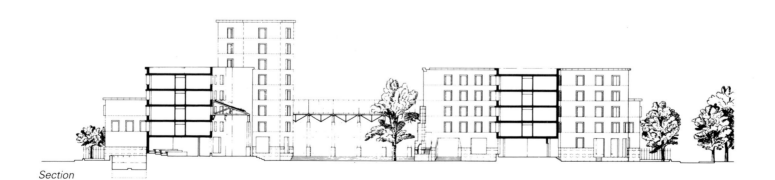

Section

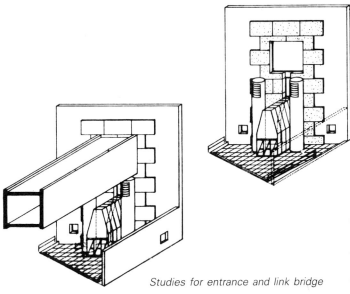

Studies for entrance and link bridge

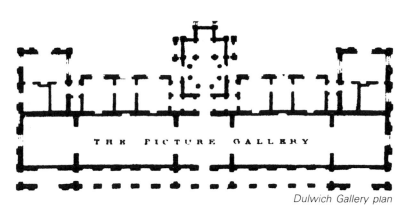

Dulwich Gallery plan

I include a plan of Soane's art gallery at Dulwich which as a nineteenth century prototype also combined smaller and larger rooms in a linear arrangement.

An article by Garry Wolf[13] dealt in equal measure with the American Academy of Arts and Sciences designed by Kallmann, McKinnell & Wood and our design for the Fogg (perhaps this collage of quotations is a device whereby I can pretend I'm not really talking about myself – maybe it's less embarrassing).

'We have lived with the villa and the palazzo for a long time now. From Pliny to Palladio, from Potsdam to Princeton, the suburban villa has offered retreat from the urban world to a contemplative natural setting... The villa we know today tends to be informal, like the life it houses. The palazzo, by contrast, is the prototypical urban building. Hemmed in where the villa is free-standing, the palazzo is recognized not by its roof but by its wall. Its architectural assignment is not to dissolve the barrier between inside and out, but to establish a strong separation between them. Where the villa turns outward to its open site, the palazzo is inward-looking. Where the one may be casual, the other is disciplined and formal. The first belongs to the garden; the second to the street. These buildings are in our memories. We recognize them readily whether they are historical monuments in Europe, or everyday structures in Cincinnati.

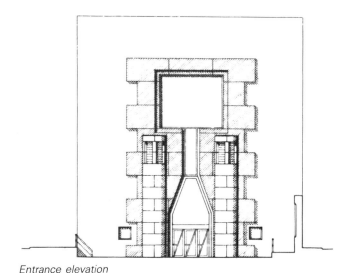

Entrance elevation

'Their very typicality may have rendered them unremarkable, while their age and tradition made them appear old-fashioned, if not obsolete, especially in the last fifty years when architects largely abandoned such prototypes. Recent decades have given Boston and Cambridge a collection of landmark structures that consciously avoid fitting into any such comfortable recesses of collective memory... It is therefore of some interest that the two latest architectural projects of note in Cambridge are a villa and a palazzo... In the Fogg project, Stirling has recovered the wall separating inside and out, and the primary opening in that wall, the front door... Where the Academy with its more open site utilized a loose assemblage of discrete rooms and open spaces around a central hall, the Fogg addition isolates and emphasizes every part of the building, without a trace of the freely flowing space of modernity... The disconnectedness of this planning serves a very different purpose from that of the Academy, for here the goal is to establish a sequence of spaces that will continue via a bridge over Broadway into the galleries of the old Fogg, to conclude in the Fogg's gracious Renaissance revival courtyard. The dramatic proportions of Stirling's vestibule and stairway establish a momentum which, with allowance for browsing along the way, will carry the visitor to the top floor and back to the front of the building to cross into the old museum. Stirling has designed a building whose culminating space is across the street, in another structure. This link across Broadway, is problematic .... however, he has designed an addition that will work with or without its bridge; no doubt he had in mind the fate of Michelangelo's plan for the Laurentian Library in Florence, a building Stirling's powerful spaces recall. There the final room, which visitors would have reached after passing through the tall vestibule and the long narrow reading room, was never constructed, leaving the existing Mannerist spaces with a brilliant, if unintended, tension.'

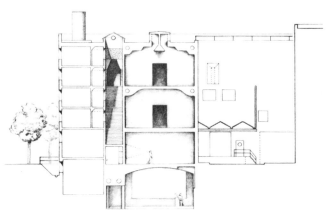

Cross-section showing the central stair with the galleries to its right

Well, I'll buy that – although I don't think I had the Laurenziana in mind – certainly not consciously. Nevertheless I'd like to think this project combines the monumental and abstract with the informal and representational.

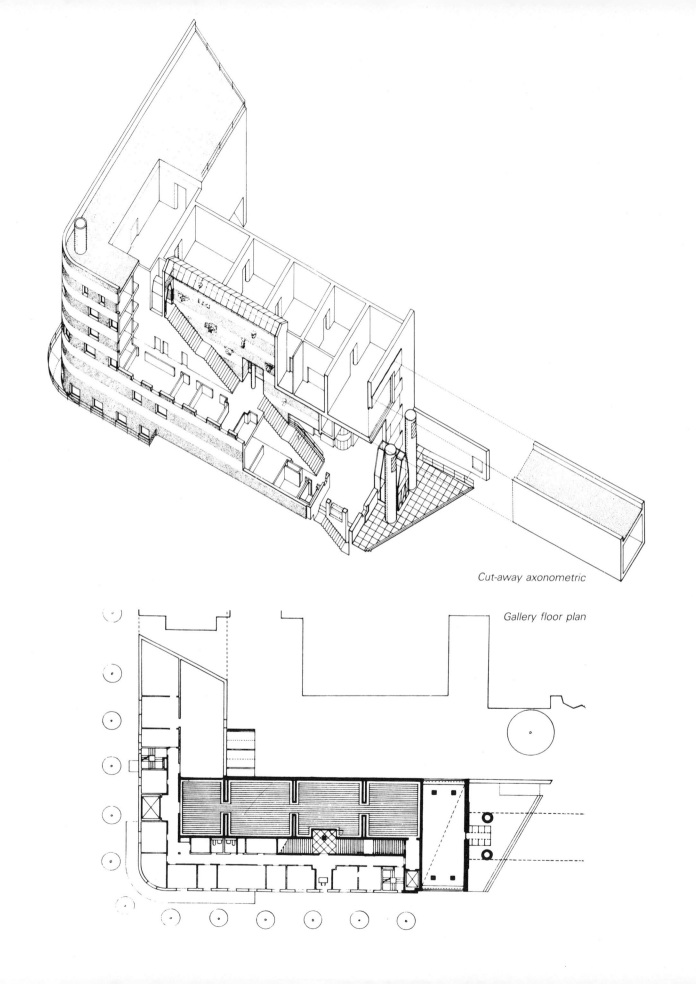

Cut-away axonometric

Gallery floor plan

Anderson Hall from the west

ANDERSON HALL, RICE UNIVERSITY, HOUSTON, TEXAS, 1979

The next L-shaped building was to double the size of the Architecture School at Rice[14] done in association with Ambrose & McEnany of Houston. The original campus by Cram, Goodhue & Ferguson[15] from the 1920s is in a sort of Venetian, Florentine, Art Deco and we were asked to work within a limited range of bricks, pantiles, pitched roofs, etc. – which we thought not unreasonable for this eccentric and elegant campus, where there are many arcades, marble balconies and fancy spires.

The existing building was L-shaped and we have extended it with another L-shaped piece. The stitching together is made by a galleria, a surgical splint of circulation, binding old and new limbs together and connecting the original entrance with a new entrance at opposite ends. Both entrances are lit through new glass spires on the roof. The galleria overlooks an exhibition space

Administration building, perspective of west elevation, 1910

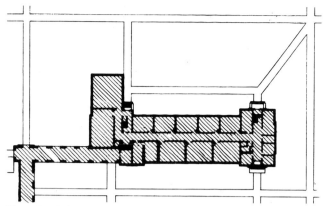

Plan of existing building

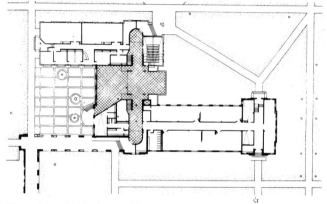

New plan with Anderson Hall added

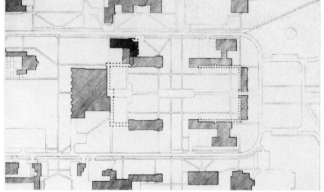

Site plan

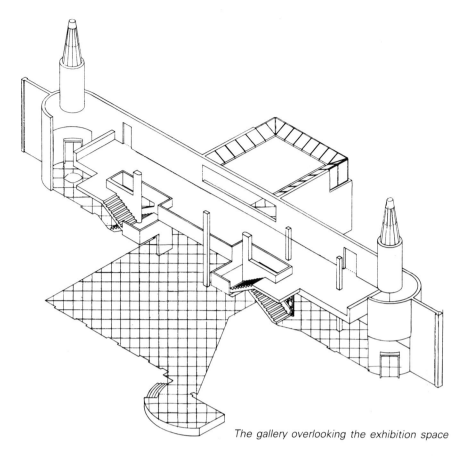

The gallery overlooking the exhibition space

on one side and a jury room on the other. The old building has a colonnade to an adjoining building and by adding a new L-shaped wing forms a three-sided courtyard, a small-scale paved garden in an otherwise open green campus.

To quote from the Inauguration Speech I had to make:

'Historically, the quality of the art in the architecture is remembered as the significant element. However, with the advent of modern architecture, sociological, functional and real estate values have come into ascendancy. Ironically, coinciding with the recent loss of certainty in modern architecture, the conviction of welfare state and hardnosed commercial standards is also, I believe, declining and there is a return to the more ancient desire for buildings whose primary objective is to appear appropriate in their context — at least that's what our clients seem to want now. Having stressed the art, perhaps I should say where I think it's at. For many of us working with the abstract language of modern architecture, Bauhaus, International Style, call it what you will, this language has become repetitive, simplistic and too narrowly confining and I, for one, welcome the passing of the revolutionary phase of the modern movement.

'I think the mainstream of architecture is usually evolutionary and, though revolutions do occur along the way (and the modern movement was certainly one) nevertheless revolutions are minority occasions. Today we can look back and regard the whole of architectural history as our background including, most certainly, the modern movement — High Tech and all. Architects have always looked back in order to move forward and we should, like painters, musicians, sculptors, be able to include ''representational'' as well as ''abstract'' elements in our art... So, freed from the burden of utopia but with increased responsibility, particularly in the urban and civic realm, we look forward to a more liberal future producing work perhaps richer in memory and association in the continuing evolution of a radical Modern Architecture. It is of course heartwarming to receive Gold Medals and Pritzker Prizes but the best prize of all is to get another building built — far more important and substantial than any award. So on behalf of Michael Wilford and myself I would like to thank all of you who have made it possible here.'

Anderson Hall, the courtyard seen through the colonnade

Broadway corner – proposed

CHANDLER NORTH, COLUMBIA UNIVERSITY, NEW YORK, 1980

Our third project in the US was to extend the Chemistry Department at Columbia University,[16] in association with Wank, Adams & Slavin of New York.

The site at the top left corner of the campus is occupied by the Gymnasium, the roof of which is an outdoor 'ground level' for this area of campus. Our problem was how to make an air rights building over without piercing the Gymnasium with new structure.

A characteristic of McKim, Mead & White's campus plan is the gaps between buildings whereby an impenetrable wall is not presented to the adjoining neighbourhood. To have merely extended the Chemistry Department with the amount of accommodation required could have resulted in a new building extending and connecting with Pupin (the Physics building) thus producing a continuous wall of building along the edge of the campus that would have been untypical and perhaps hostile to the neighbourhood.

When the Gymnasium was constructed four heavy engineering columns were positioned to allow an air rights possibility. We made use of these and also found we could get new structure down along the Broadway boundary and into the service yard behind the Chemistry building. This allowed us to design a bridge-type structure which slews half the floor plan diagonally back into the campus and thus maintains a gap between the end of the new extension and Pupin. This diagonal wing reduces the amount of outdoor terrace in this part of the campus where it is in any case too extensive and under-used. The new extension combines the masonry appearance of McKim, Mead & White's buildings, particularly along the Broadway frontage, with an engineering appearance related to spanning the Gymnasium.

Accommodation is mainly laboratories, instrument rooms and staff offices; however, at the junction of the diagonal and regular wings and overlooking the gap there is a social lounge/ reading room which has views inwards across the campus and outwards towards Riverside. I would like to regard the project as a balanced juxtaposition of abstract/High Tech with representational/traditional.

I hope I have not focused too narrowly on a single aspect of our work, but it does seem to me that, if one is not over-concerned with the desire for stylistic consistency and personal authorship, there does exist the freedom to respond – or not – to the context of a site and to the appropriateness of a situation.[17]

Campus plan – the Chemistry Department is top left

Laboratory floor plan

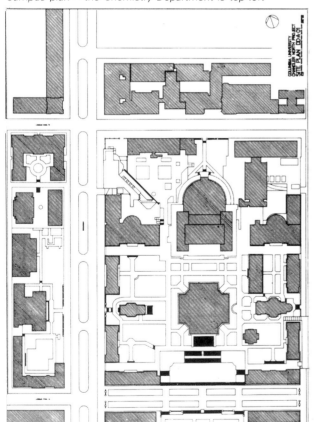

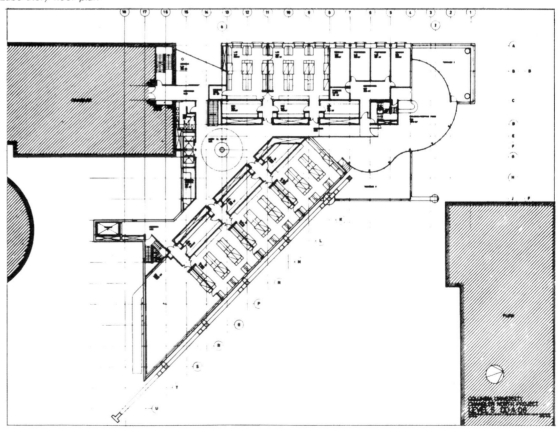

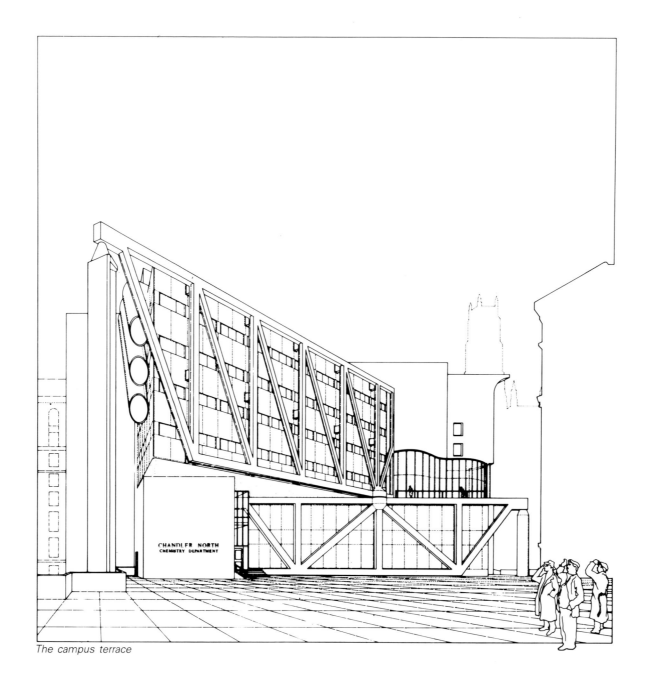

The campus terrace

NOTES

Mavrolean House, 1957 and Leicester Engineering Building, 1959 were by James Stirling and James Gowan. St Andrews Arts Centre, 1971; Düsseldorf Museum, 1975; Stuttgart Museum, 1977; Wissenschaftszentrum Berlin, 1979; Extension to the Fogg Museum, 1979; Anderson Hall, Rice University, 1979; Clore Gallery, Tate Gallery 1980; and Chandler North, Columbia University, 1980 were by James Stirling and Michael Wilford.
All other projects were by James Stirling.

1 This (slide) talk was first given at Yale School of Architecture on 3 November, 1981 and the following day at Columbia University, New York. Then at the Architectural Association in London, also in Paris and Cornell University, USA. In April 1982 at the Graham Foundation, Chicago and the following day at the Architecture School in Champagne Urbana. Then in Europe; Bergamo, Munich, Darmstadt and the American Academy in Rome. Usually I have a slide show related to our current work and I try to make a new version about every two years.

2 *Architectural Design* Vol. 49, No. 3/4, 1979, also *Roma Interrotta, Incontri Internazionali D'Arte*, Rome, Mercati di Traiano, May/June 1978.

3 For all projects until 1974 see *James Stirling, Buildings and Projects 1950–1974* published by Verlag Hatje in Germany (also Thames & Hudson, UK and Oxford University Press, USA).

4 *Architecture + Urbanism*, July 1976, also *9 Architects*, Catalogue, Dortmunder Architektur-ausstellung 1976, *et al.*

5 *Architectural Design* Vol. 49, No. 3/4, 1979.

6 *Architectural Design* Vol. 49, No. 8/9, 1979.

7 See letter from M. M. Reichert and H. Reichert in *Architectural Design* 9/10, 1977.

8 *Architectural Design* Vol. 50, No. 7/8, 1980, *et al.*

9 *Architectural Design* Vol. 52, No. 1/2, 1982.

10 July 15, 1981.

11 *Architectural Design* Vol. 52, No. 1/2, 1982.

12 *The Architectural Review* No. 1011, May 1981, also *Casabella*, May 1982, *et al.*

13 *The New Boston Review*, October 1981.

14 *Progressive Architecture*, October 1980, also *The Architectural Review* February 1982, *et al.*

15 See *Architecture at Rice*, Monograph 29 by Stephen Fox.

16 *The Architectural Review*, December 1982.

17 With the first talk at Yale I concluded, 'And in a special sense I would like to thank you — the students and faculty here — who by continuously asking me to teach at Yale have in effect provided me with a base in America.'

Jørn UTZON

THE IMPORTANCE OF ARCHITECTS

Human beings experience their surroundings in different degrees. If you have an extreme sensitivity for the impact of the light and shapes, colour and space which are surrounding you, you have the inborn qualities of an architect and an artist.

If you are not just receptive but also have a creative talent and are able to express yourself so that your experience can be understood and enjoyed by your fellow-men, then you have some of the qualities necessary for becoming an architect, an artist. Art is the liberation of the creative forces within you.

Recently I asked the Danish painter, Egil Jakobsen, *What is Art?* He answered that, when he was an active member of the COBRA-group consisting of marvellous spontaneous painters like Asger Jørn and Dubuffet, they had been discussing this very question and their conclusion was that they considered Art as the result of the release of the inner creative forces in man, in you. A Swedish poet speaking of Art issued this warning: *Do not allow the intelligence to stand in the way or block your feelings from passing through the exit-door.*

The artist/architect Louis Kahn regarded the university as the place for learning and the place for the creation of new ideas. He defined the university as the threshold between light and darkness. In the light he placed science and knowledge, everything here fully exposed and totally exact, everything here proved and known. In the darkness he placed ideals, dreams, aspirations, feelings, imagination, intuition – all possibilities are open, you are in the sphere of the unknown. On the threshold – Kahn's university – intuition and knowledge meet and work together so that totally new things are created.

Architecture – in a similar way – is based on science as well as on intuition, and if you want to become an architect, you will have to master technology in order to develop your ideas, in order to prove that your intuition was right, in order to build your dreams.

One of the originators of the Atomic Age, the scientist Niels Bohr – during a discussion of the importance of the science of mathematics – expressed the opinion that mathematics was only a tool with which to prove what you had already discovered and established by intuition.

You can study architecture from many angles, from an historic point of view for instance. You can sort out buildings belonging to different periods and – therefore – built in different styles: Renaissance, Baroque, Rococo, etc., but you can also – as the Italian structural engineer, Luigi Nervi – evaluate buildings according to the structure and the building technique applied. Luigi Nervi has reached the conclusion that the buildings considered most outstanding of a certain time and style are almost always constructed with the most exquisite building technique of that special time.

If you look at architecture in yet another way –

evaluating a building purely from the sensation of joy it gives you – you experience the building with your senses only and you become a user of the building in the way the architect had conceived it. Then you are in close contact with what the architect was aiming at.

Among the finest examples in Scandinavian architecture which make you sense how much devotion the architect has given to your well-being are two buildings by Gunnar Asplund, the Forest Crematorium in Stockholm and his Courthouse in Gothenburg. Asplund was the father of modern Scandinavian architecture. He went beyond the mere functional approach and created a wonderful feeling of well-being in his buildings. He even added a symbolic content which gives each of his buildings a unique personality which very strongly radiates the purpose of the building, completely covering and expressing the function and life-style, the form of life going on in the building.

In his Forest Crematorium, a group of small chapels situated in a clearing in a typical Stockholm pine forest, he leads you through an intimate waiting room across a small isolated courtyard and takes the mourners one by one through a small door into the chapel. The chapel itself is rather dark, but after the ceremony the rear wall disappears, daylight pours in and the mourners leave the chapel together for a short stop under the big roof of the open hall, which faces a very simple, silent landscape, just a treeless, grass-covered hill meeting the sky. Here, in the middle of the forest, where the view is normally always limited by trees, this gives a strong feeling of peace and eternity. The chapels and the forest-edge are married together, white marble buildings among evergreen pine trees.

Gothenburg Courthouse is an extension of an existing townhall. The main space of the building is a hall through two storeys with balconies – full of light, flowers and plants – with walls of light natural wood, with very comfortable furniture and with a refined detailing. This is the waiting hall for a number of courtrooms, and it gives you a feeling of friendliness, warmth and purity. It stimulates expectations of justice and understanding, not just of punishment. It has an atmosphere quite contrary to – for instance – that of the main Courthouse in Copenhagen which – with a sinister closed elevation with heavy columns and dark spaces – is an awe-inspiring building which seems to equate Law with Punishment.

I have now told you about some of the ingredients which go into the foundation on which I build my own architecture, and here follow three projects:
A House of Festivals (Sydney Opera House)
A House for Work and Decisions (Kuwait National Assembly Complex) and
A House for Family Life (Vacation House in Spain).

Forest Crematorium, Gunnar Asplund

Sydney Opera House
*Concept sketches: the Japanese house of roof and platform; clouds over
the horizontal sea; preliminary sketches for the vaults and later studies*

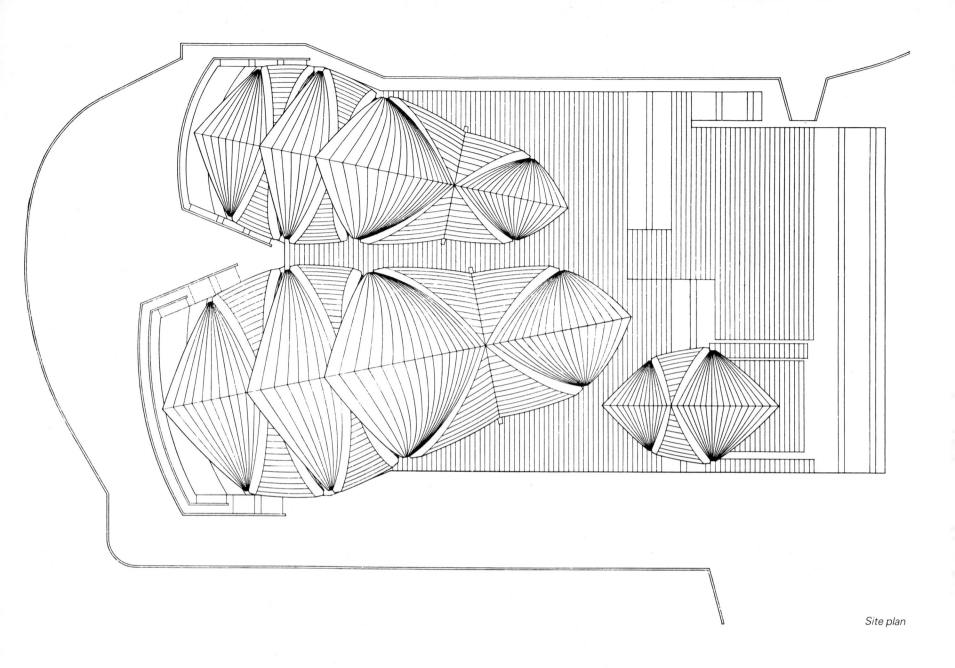

Site plan

A House of Festivals: Sydney Opera House

The situation on the peninsula in the beautiful harbour makes Sydney Opera House a focal point in Sydney, because big liners, ferry-boats and yachts sail around the peninsula and because you look down on it from the harbour bridge, from the skyscrapers nearby and from the surrounding town built on the hills of Sydney. The Opera House is completely exposed and is therefore one of those buildings where the roof is of major importance — one could not have a flat roof filled with ventilation pipes — in fact, one must have a fifth facade which is just as important as the other four. This is why — instead of covering the various functions of the building under one big square box — I have expressed all the different spaces within the building by covering them by a group of shells; I have made a sculpture of it. The interplay of sun, light and clouds makes a living thing out of this sculpture, and in order to express this liveliness the shells are covered with glazed, shiny, white tiles.

The organization of the building has grown out of this particular situation in a natural way. The whole peninsula is covered with a closed, almost rock-like building, the base, housing the artists' isolated world for their concentration and preparation.

On top of this base is a grand, stepped platform with two amphitheatres with stagetowers and foyers lying side by side, each of them covered with its own group of great white shells.

Here — on the platform — is the meeting place between the artist and the audience, the stage openings; here the spectator receives the artist's final product.

The separation between artist and spectator is ideal — the spectator is where he is meant to be, on top of the platform, in close contact with the majestic harbour landscape, so different from his daily world, so ideally suited to be the overture to the time of imagination and surprise he has come for.

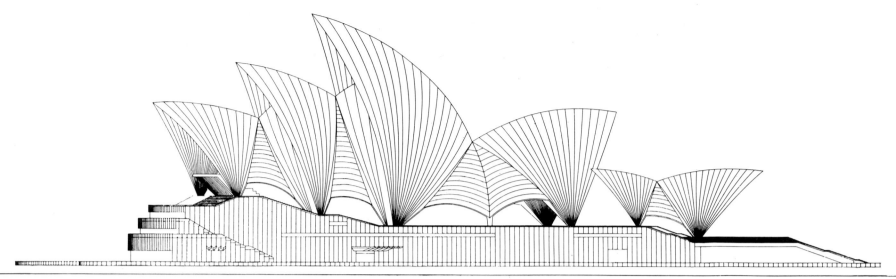

West elevation

South elevation

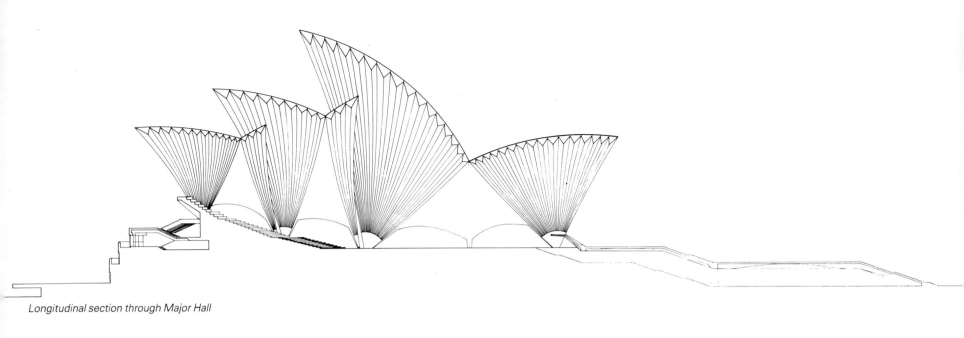

Longitudinal section through Major Hall

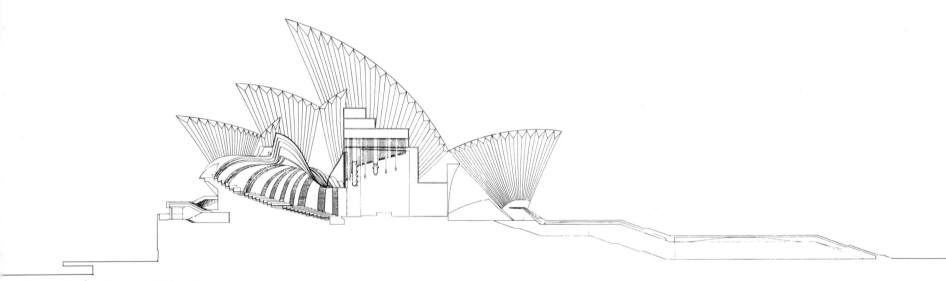

Longitudinal section through Minor Hall

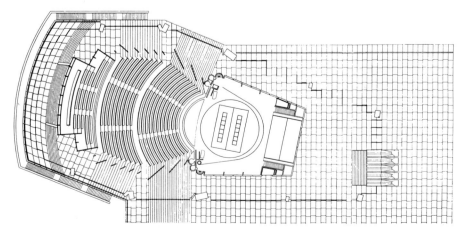

Plan of Minor Hall

A House for Work and Decisions: Kuwait National Assembly Complex

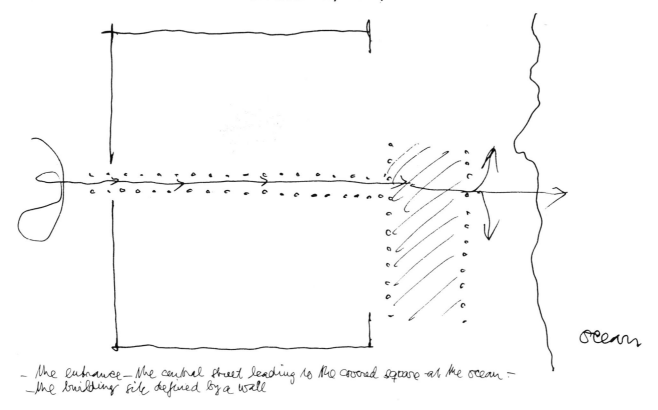

— the entrance — the central street leading to the covered square at the ocean —
— the building site defined by a wall

All departments of the complex (offices, meeting rooms, reception rooms, library, Assembly Hall, etc.) are arranged along a central street. The departments consist of modules of various sizes, built around small patios or courtyards, connected to the central street via side streets. Each department can be extended at any time by adding modules, so that the building can grow sideways, away from the central street, and its outer boundaries will change as time goes by. These free-flexing outer boundaries of the system are very much related to traditional Islamic bazaar architecture.

The construction of the National Assembly also reflects the purity of Islamic construction. The building is a prefabricated concrete structure in which all elements are structurally designed to express the load they are carrying, the space they are covering — there are different elements for different spaces. They are all meant to be left visible — contrary to the constructions of the 'cardboard-architecture' of most modern office and administration buildings where hidden structures, lowered ceilings and Gypson walls give you an impression of being in a cardboard box.

In the National Assembly complex you see very clearly, what is carrying and what is being carried. You get the secure feeling of something built — not just designed.

The demand for very busy intercommunication between the various departments has led to the decision to arrange the complex as a two-storey building. This provides an easy orientation inside the building in contrast to the abrupt disorientated feeling you may experience in buildings with many floors with intercommunication depending on elevators.

When you enter the central street, you can see all the entrances to the various departments. The orientation is as simple as the orientation you get when you open a book on the first page with its table of contents presenting the headings of all the chapters.

The central street leads toward the ocean into a great open hall which gives shade to a big open square, where the people can meet their ruler. In Arab countries there is a tradition for very direct and close contact between the ruler and his people.

The dangerously strong sunshine in Kuwait makes it necessary to protect yourself in the shade — the shade is vital for your existence — and this hall which provides shade for the public meetings could perhaps be considered symbolic for the protection a ruler extends to his people. There is an Arab saying: 'When a ruler dies, his shadow is lost'.

This big open hall, the covered square, between the compact closed building and the sea, has grown out of this very special situation in quite a natural way — caused by the building's position directly on the beach. This big open hall connects the complex completely to the site and creates a feeling that the building is an inseparable part of the landscape, a feeling that it has always been there. The hall is just as much part of the openness of the ocean as it is part of the compact building and its structure. The hall seems born by the meeting between the ocean and the building in the same natural way as the surf is born by the meeting of the ocean and the beach — an inseparable part of both.

The Kuwait National Assembly project is at the time of writing still under construction. It has reached the stage in between the start — when the architect is facing a white sheet of paper — and the finest moment — when the echo of the last stroke of a hammer has died out in his building — and the opening ceremony can take place.

After a period of intense co-operation between architect, structural engineers and contractors this middle stage is especially delightful for the architect: activities all over the place, expert craftsmen handling equipment with greatest precision dealing with elements weighing up to several hundred tons, space being encased and brought into existence — for the first time the architect is able to walk in the spaces which until now he could only house in his imagination.

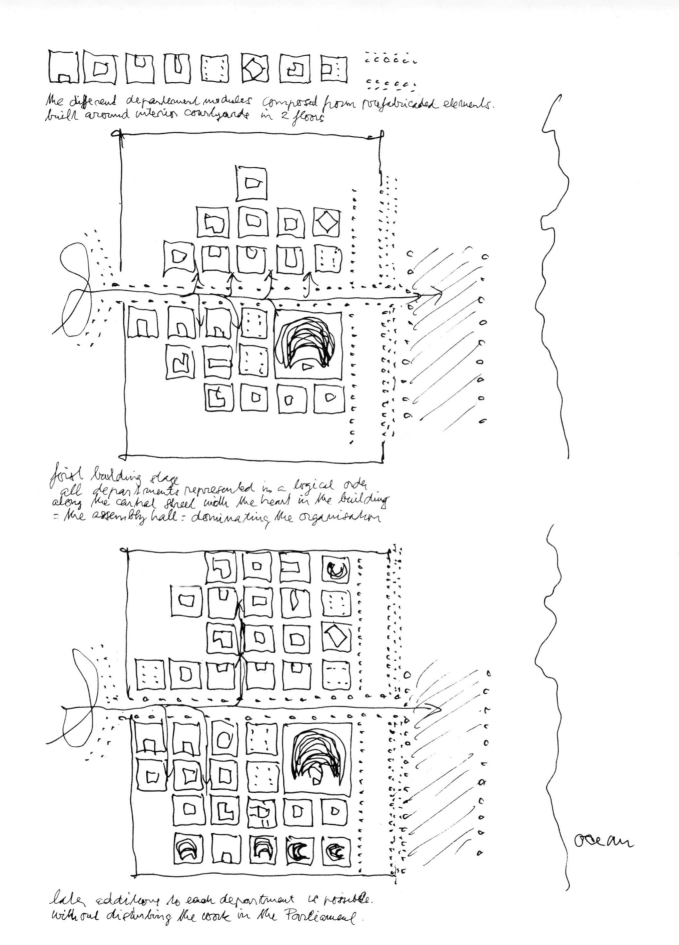

The different department modules composed from prefabricated elements, built around interior courtyards in 2 floors

first building stage
all departments represented in a logical order
along the central street with the heart in the building
= the assembly hall: dominating the organisation

later additions to each department is possible,
without disturbing the work in the Parliament.

ocean

A House for Family Life:
Vacation House in Spain

It is a sandstone house on the edge of the cliffs twenty metres above the Mediterranean, built with stones cut out from the cliffs.

It houses only one single room, totally dominated by one big, curved couch which embraces the whole family.

The deep window-wall softens the glare from the sun and the sea. The window frames are outside — invisible from inside — so you are alone with sandstone, sky and sea.

A narrow slot in the west wall invites the sun for a visit to the south wall for a few minutes every day, making you aware of the passing of time.

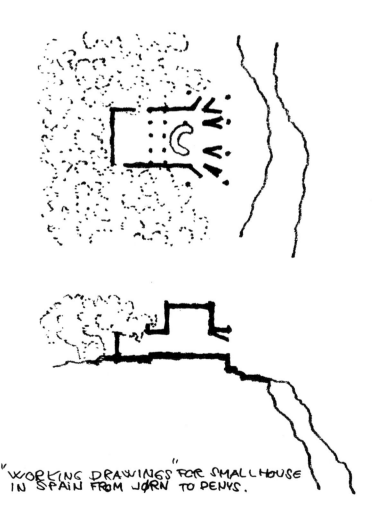

"WORKING DRAWINGS" FOR SMALL HOUSE IN SPAIN FROM JØRN TO PENYS.

WORKING METHODS

The working methods of architects differ widely. The human brain is constructed in such a way that it can – simultaneously – put forward solutions to problems and evaluate these solutions. Creative architectural work is of such a complicated nature that in order to develop satisfactory results an enormous amount of drawings and models is necessary.

A greater part of this work may seem wasted – in the process of purifying and intertwining all the different parts and functions of a building you usually work with so many alternatives (in order to find the right one) that as much as ninety per cent of your work might be thrown away. Instead of complaining of this waste you can compare it to the wasteful abundance in nature.

Le Corbusier has expressed this in a comforting way: *the architect's work is never lost – the work with each building has something in it for the next one.*

The architect becomes increasingly professional.

Le Corbusier was a professional and a revolutionary. He believed in the possibilities of his own time, he believed that in order to be true you should be creative with the means of your own time for your own time.

An example which reveals this bold attitude of his is his remarkable statement, when he had been visiting the Sistine Chapel: Le Corbusier and the Italian architect Gio Ponti walked on the scaffolding during a period of restoration of the chapel and were for several hours at only arm's length from the frescoes of Michelangelo. They walked in complete silence and, when they left the chapel together, Gio Ponti was anxious to hear what Le Corbusier would say. Then came this amazing statement: *Give a good man, a real artist, a brush and some colours, let him work undisturbed for five years, and we shall have a work just as important!*

CONCLUSION

Of all persons involved in the building process, the architect is the only one whose aim is to create the most ideal conditions for human beings out of the programme and the means given to him. The other participants each have a different niche: the engineers seek to achieve the ideal performance from equipment and stability from structure, the contractors get the building up and are responsible for the actual construction, the financiers and the lawyers are in control of the economic aspect and the client provides the programme with the basic requirements.

I finally want to clarify this special position of the architect, I want to give you an extra angle from which to see it and to throw a sharp spot-light on the issue, to make this all-important architectural aim stand out in clear relief with deep shadows. I shall let Ralph Erskine, an architect of an extraordinarily vivid and lively human architecture, sum it up. He states: *In the development of a project the client (i.e. the future user of the building with his special life style) is just as important a building material as concrete, brick, stone, timber and steel.*

Bagsvaerd Church[1]

My first impression of Bagsvaerd Church in the bright morning light was of a building whose presence recalled fundamental things about a meeting house. It is an unselfconscious, direct, concrete frame and pre-cast panel job – white, immaculate, and light in feeling. It is a long building on plan, 260 feet by 70 feet, with a remarkable section typical of Utzon's work.

The long outer boundaries of the building each consist of two parallel lines of precast concrete frames on an 8 foot module. They are clad with white concrete slabs and are set some 56 feet apart. Gradually, from the east, they step and build up (like extended Flemish gables) to a peak at the west end where, some 50 feet high, they replace traditional buttresses. They support and provide Utzon with the freedom to span between with thin-sprayed concrete shells (like folded, free-form vaults) over the main body of the church. I find the operation of these two opposing systems (the regular module and the free-form vault) positively life enhancing.

Light diffuses over the white and grey surfaces of the interior of the vaults, the galleries, the elements of the altar, and the white concrete flagstones of the floor.

Clarity of structure, scrupulous attention to detail all express wih affection the inner rhythms and pulse of the design. The whole thing flows out of its own order. It is a contemplative space much loved by the community it serves and comparable to Corbusier's chapel at Ronchamp.

1   Extract from Denys Lasdun's speech on the occasion of the presentation of the 1978 RIBA Gold Medal to Jørn Utzon on 20 June.

Concept sketches

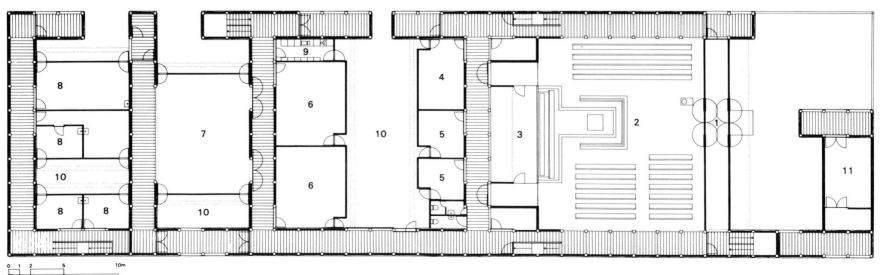

Plan

1 Entrance 2 Church 3 Sacristy 4 Waiting room 5 Office 6 Candidates' room
7 Parish Hall 8 Meeting rooms 9 Kitchen 10 Atrium garden 11 Chapel

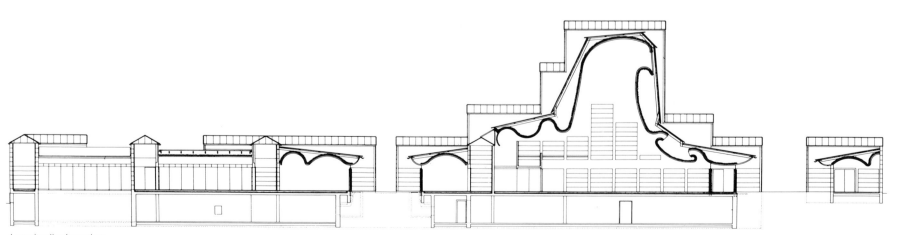

Longitudinal section

233

Aldo VAN EYCK

WASTED GAIN

Man's scope – our scope – spans disquieting extremes. The same can be said of his tools. Maid or master, technology assists all our doings at every level of intention. Constructive or destructive, it is always by our side – a kind or malicious companion, both.

The evidence of this companionship can be read off the face of the globe, for environment reveals whatever occurs there – like a tell-tale mirror. What we see (if we dare to look) reflected in the mirror – the evidence – is becoming more and more harrowing – even pathological.[1] Indeed, it is befitting today to talk of the pathology of landscape.[2] Of course you can nod and say: 'Yes, we know, we know!' But **do** you know – has it really penetrated your consciousness, or are you willing to acknowledge that, whilst in the past societies responded more or less successfully to the problems survival in any given environment poses, **ours**, those to which you and I belong – the magnanimous ones – are no longer able to? This in spite of bewildering technological and scientific ability – their familiar trademark.

But it is another trademark – a less familiar one now edging in on us – which I want to put before you. I mean our pitiful inability to come to terms with – cope with – vast multiplicity and great quanta no matter of what – and to behave with sanity towards environment, that great and only place there is in which we all live.

No previous society has made quite so little of the experience, knowledge and technology available, or fallen so far short of what imaginative concerted action could bring about. Of course, as of old, there are still imaginative and constructive efforts by individuals from which all could benefit if listened to and applied. But minimal use is made of them. In fact, a blind eye and a deaf ear is turned to what is being done on the periphery – to what is occasionally smuggled in like contraband – only to be misunderstood and misused, twisted into yet another negative.

Whatever gain is made is soon counteracted by another gain, for ours – yours and mine, the presumptuous kind – is a society of wasted gain. Just think of it. Ten years ago, when war was still raging beyond the Pacific, I wrote some lines in big white letters on an exhibition wall in Milan.[3]

Here they are: 'Disturb the delicate intricacies of the **limitless microcosm**. Trespass into the **limitless microcosm** and frighten the angels. In between mess up the Mississippi and the Mekong. If that is what we desire, it will soon be here, for there is a limit even to the limitless. Man falling in line with entropy after all. We have already turned the theory of relativity and quantum physics against ourselves. We now split atoms – we split and split. Soon it will be the stars. **From limited total loss to limitless total loss. Mourn also for all butterflies.**'

Since then, ten years have passed. What is true of the whole is also true of each part, so what is true – painfully true – of environment is also true –

---

1   Key address delivered for the sixty-third commencement of the New Jersey Institute of Technology, 1979.
2   A term coined by Piotr Gongrijp, a Dutch architect and environmentalist.

painfully so – of most buildings. Now if making a **good** building (one that is not a **bad** building) has become too difficult, the dilemma is indeed complete. But is it really all that difficult? Does it really require a genius to avoid the mean and meaningless – or a sage to bypass foolishness? Is there nothing between a fool and a genius – nobody in between to do the job nicely – well?

If behaving with sanity towards environment on whatever dimensional level is no longer within our reach – within reach of the kind of society to which we belong – then surely such societies – ours – thus reasonably gauged, are of a low – primitive – order.

I have just confronted you with a fact which, alas, points towards the only definition of a 'primitive' society that makes sense today: a definition which removes the burden from all those **other** societies; the kind that never really deserved to be called primitive by that self-assured kind of ours, which behaves towards the landscapes of the world like a half-wit with two left hands.

Whether in Greenland, Africa, America-long-ago or the South Seas, people dealt with limited numbers both accurately and gracefully, extending collective behaviour into adequate and often beautiful built form. Taking from environment as much as they gave, a gratifying balance was sustained. This **we** are no longer able to do, not in the same way, nor as yet, in any other way.

Yes, is it really all that difficult or beyond our reach? Do we not belong to the same species as those in Greenland, Africa, America-long-ago and the South Seas? Is our mental equipment not similar? Are we not endowed equally well? Surely we can accomplish what they accomplished in so many different ways. Believe it or not, those little societies and their fast-disappearing cultures can by their example still tell us that abilities (hence also possibilities) which we have come to regard as beyond the scope of the human being **actually lie within it**. And so there is no reason to give up all hope.

TRIENNALE DI MILANO, 1968: IL GRANDE NUMERO
Plan with building directions, captions on the wall and descriptions of the exhibition material

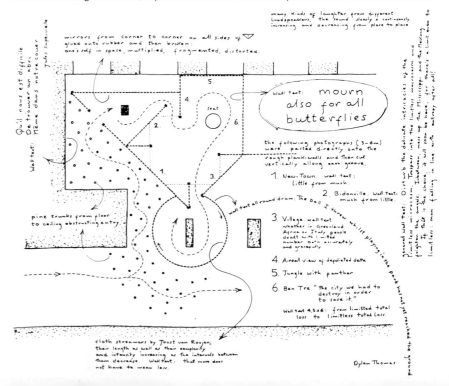

For it is in the nature of the human species – **all** people, you see – to be able to deal with environment, hence also to fashion the spaces they require, adequately – and sometimes beautifully. The way people are also given to communicate with each other through language – speech – **that other gift** which, like making spaces, still belongs to their primordial equipment.

It is painfully true of architecture that it is not just good quality that counts but a sufficient quantity of that quality. A good school elsewhere is no use to a child in need of one here. If I have a nice house (a house at all), it does not mean that millions upon millions of others also have. So let's get moving and start with this in mind: **Today the architect is the ally of every man or no man**. Persuade those narrow borderlines – the hard and harsh ones – between inside and outside, between this space here and that space there – to loop generously and gracefully into articulated in-between places and give each space the right interior horizon for the gratifying sense of reference it provides. Never cease to identify whatever you construct with the people you are constructing it for – for those it will accommodate. Identify a building with that same building entered – hence with those it shelters – and define space – each space built –.simply as the appreciation of it. This circular definition has a purpose, for, whilst it excludes all abstract academic abracadabra, **it includes** what should never be excluded but paradoxically generally is: I mean those **entering** it, appreciating it – **PEOPLE**.

Architecture can do no more, nor should it ever do less, than accommodate people well; assist their homecoming. The rest – those signs and symbols one is worrying about far too much – will either take care of themselves or they just don't matter.

Meanwhile, let us not forget that blight has crept over our field. Keep clear of the entire array of current whimsy-flimsy trends; keep them from nestling in your minds. And do whatever you can to prevent those concerned from being tricked into actually building the vicious soft-coloured absurdities that fill most of today's architectural reviews. Architecture does **not** mean, nor has it ever meant, nor will it ever come to pass that it could one day mean,

3
Huge halls built by order of Mussolini were subsequently painted a sick brown in their entirety so that all contributions would stand out in hopeful contrast. Since, however, the general subject was 'Greater Number', all walls of this particular contribution (unplaned planks five metres high from floor to ceiling) were painted the same sick brown! The structure – a set of weird, outsized crates – contained uncomfortable items and was intentionally hardly accessible.
Ironically the entire Triennale was occupied by dissatisfied artists immediately after the official opening and thus remained literally inaccessible.

**Identify a building
with that same building
entered, and space simply
as the appreciation of it**...

what the loathsome fives, **sixes** and sevens of New York, **Minnesota** and Chicago or their like are trying to make people believe it means. So beware especially of New York and London and Milano – the worst trend head-quarters – where they are still busy bending over backwards trying to twist our profession into something it simply is not. Not even in the sense that an apple is not a pear but still a fruit. That architecture – buildings – should no longer help mitigate inner stress, but should, instead, provoke it, is hardly a conceivable objective. And yet here it is, flourishing on both sides of both great oceans, escorted by sickening flirtations with absurdity, irony, banality, inconsistency and, of course with ROME, Rome and **Rome** again –.the most obnoxious pests afflicting architecture since Fascist gigantism and Nazi blood-and-earth regionalism in Europe forty years ago. It is worth noting that the new historicists and eclectics, whose habit it is to misquote the past, instead of coming up with a large variety of cocktails, produce – all of them together – little more than a single standard watery monomix. So never mind the Minnesota Six! **As for history, that wonderful body of gathering experience, it is there to help – NOT TO BE SPILT.**

What is needed is better functioning – on more levels this time. Just that. For there is no such thing as a solid teapot that also pours tea. Such an object might be a penetrating statement about something (and thus perhaps a work of art), but it is simply NOT A TEAPOT since it CANNOT POUR TEA. Neither is there – nor will there ever be – such a thing as a building which is **wilfully** either absurd, trivial, incoherent, contradictory or disconcerting, that is still a building. Marcel Duchamp invented puzzling objects but they were significantly **not** buildings!

It will not be long before the earth's face will be like a network of scars. Energy too is spilt and ebbing. Time is ticking faster. Millions have no place to go – no nothing. What **can** be done that is more effective than trying to save the world? What could an Institute of Technology like this one, for example, do as soon as it is ready to do so? Well, start dissociating technology – setting it free – from that ruthless and naive notion of progress to which it has been falsely tied for so long: For progress means nothing on this side of evil if it does not mean moving towards well-being for **all** people – (and all people means simply that – **all** people) – and away, to start with, from quite so much waste, pollution, discrimination and unnecessary poverty. I am aware of the fact that many of those present have, over the years, sacrificed a lot in order that a limited number could study and thus achieve what was still out of reach in their time: scope and space in which to move forward. It would be gratifying if more than a few of those graduating now would in turn wish to share the scope acquired with others not yet as fortunate so that they too may move.[4]

It would, of course, also be gratifying if, in a future not very distant, those graduating on a morning such as this would represent the magnificent diversity of your country's people rather more accurately.

I can think of no single word as appropriate and sparkling as the one with which I shall now end: solidarity.

4   Unlike other institutes of higher education in USA a large proportion of the NJIT student body has a working class background.

235

THE DECLARATION OF DELFT!

THE PRIORITY JOSTLE[5]

A lot has been said here about countless people in places everywhere. That is what happens when 'experts' meet to debate the problems of others. It is also what one can expect. But what I did not expect, although I might well have, is that the nature of what was brought forward – the assortment of arguments, solutions, and strategies – is such that I can only hope that none of it will leak from this auditorium.

The question why vast multitudes migrate towards cities from rural areas is posed again and again. Many alternative reasons for this universal phenomenon are given as though its very universality points to a single set of reasons or the 'same' reasons must account for vast multitudes doing the same thing! We have just listened to the usual spate of numerical acrobatics; to dark consequences coolly outlined and redeeming solutions smoothly proposed. 'If nothing is done . . .' or, more ominous still: 'If nothing is done to stop them. . .' That is the music murmured. Stop whom? how? why?

Yes, why? Why stop them? 'It so happens,' Dr Königsberger, who knew what he was talking about, reminded the meeting, 'that nothing can stop them and nothing should.' That saved the day, although it was probably not what one wished to hear. Never mind: if the causes are elusive, the solutions may perhaps also be! But that soon proved to be a disappointment. Different *experts* proposed quite different *first* priorities. To mention just a few: one is a roof and its supports. But what if some people somewhere have reasons to prefer a walled enclosure (without a roof) for lateral protection first or, like many a Paris *clochard*, simply don't seem to want a roof in the first place? Another expert's *first* is running water and street lights; then, once the house (presumably a *second* priority) is built, also light inside! This, the message stresses, would counteract birth explosion more effectively than the pill! There were, of course some more first, second and third priorities less impertinent brought forward by other experts, but let me stick to the *firsts* I have just mentioned (roof-support and water-light) and look at Lima for a moment.

What if a poor Limeño family, escaping from the burden and oppression of urban slum life, desires to build a house of its own and starts the process, as so many do, by erecting an enclosure – just four walls – to protect, among other things, the material with which to build that very house? As for running water and light, the Limeños leave both behind them voluntarily when they evacuate

from the city and go to a barriada to follow their own priorities: a house which is theirs with lots of rooms ultimately on land from which they hope they will not be evicted – and a better life and education for their children which, incidentally, implies a proper pair of shoes first. You see, for the sake of that school – often self-built – entered wearing those little shoes (without which no parents would ever send a child to school) **some people are willing to leave running water and electricity behind them**. These people establish or join a community willing and able to defend its **own** first preliminary rights and priorities. One of the latter may be – take heed over there – to erect a nice facade with little behind it. To *show off* perhaps, which could well mean to suggest – anticipate – the presence (possession) of what is not yet there, but, by God, one day will be. An *expert* in these matters back from Nigeria has just explained why there is but one modern building in downtown Lagos with timber windows – the American Embassy. 'Africans, you see, prefer steel, though it has to be imported and rusts quickly.' I did hear the burst of laughter that atrocious remark evoked – **but missed the joke**.

We are often reminded of the United Nations Declaration of Human Rights – the one about each man's inalienable right to food, health, clothes and shelter. I have probably missed some of the great goods proposed during this meeting and got the sequence wrong. Anyway, I should like to propose another, simpler one and call it the Declaration of Delft! It affirms the right of every society to its **own** priorities as well as the right of each individual to his or hers. Just that; no more.

Ultimately the condition of the human species cannot be what a small portion of that species has so amply measured out for itself – nor is it necessary or desirable. The discovery of another standard is yet to come: it may be a hundred, it may be hundreds or more years hence, but, sooner or later, people will accommodate themselves to its nature and respond intelligently – tune in – to its prerogatives. It will not be a 'lower' standard, nor will it be a 'higher' one. It will be a **different** one: an altogether **different kind** of standard. **Whilst not everything desirable is possible, nor everything possible desirable – ecologically speaking, only the possible is desirable**. Hence there is still lots of scope for quality which, unlike quantity, is unlimited – boundless – and calls for a different kind – or different use – of intelligence.

Whose problems are you trying to solve if they are not your own? Who calls upon a few in the name of all? The Lord? All? or just you?

5 Introducing the Barriadas at a symposium about building in the developing countries of the tropics held in Delft, Autumn 1970.

eighteen square metres per family...

running water...

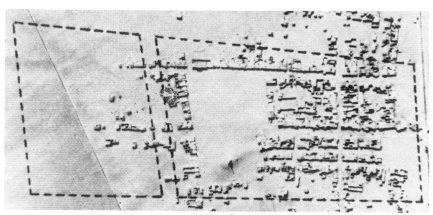

The site (dotted line) superimposed onto an aerial view of Ciudad de Dios barriada

Footnote! WHO ARE WE BUILDING FOR AND WHY?[6]

The answer to the first question is clearer in Lima than it is elsewhere in the world, since Limeños are neither mute as to their aspirations nor are they passive with regard to effecting them step-by-step. What they wish is implicitly and explicitly demonstrated by what they actually do. The barriadas offer an emphatic testimony.

The question why one should build in the way implied by the PREVI programme is another and extremely critical issue from many aspects – economic, social and political as well as architectural. There are, of course, some positive reasons why one should. However, less positive implications, which may very well be lurking round the corner, must constantly be kept in mind.

It would, for example, be a grave error if pre-designed and partially pre-constructed urban environments such as this pilot project proposes should counteract the growth and development of the barriada idea and practice, instead of stimulating it through the erection of improved dwelling types, construction systems and overall community planning.

The needs and aspirations of the people are revealed in barriadas like Comas, San Martin de Porres and Ciudad de Dios, as well as in partially preconstructed settlements like Ventinilla and Pamploña. Each are, of course, different in their possibilities and lack of possibilities, yet basically there is little or nothing to show that initially people who will buy and extend a dwelling in a pre-designed and partially constructed settlement are different or have substantially different aspirations from those who go to the barriadas to build from scratch both their own house and the community they themselves have initiated.

6 This anticipatory warning critical 'footnote' went with our competition entry – see *Architectural Design* No. 4, 1970.

Pampa de Comas barriada, Lima

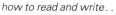
electricity...

how to read and write...

LOW-COST HOUSING IN LIMA, PERU

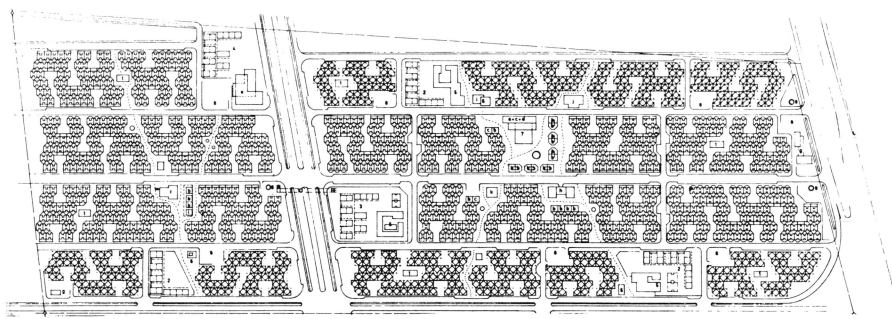

Site plan – busy 'straightforward' avenidas for both pedestrians and cars

In 1966 the Peruvian Government, assisted by the United Nations Development Programme, formulated an Experimental Housing Project (PREVI) whose aim was that of developing, through foreign and Peruvian contributions, methods and techniques to be applied afterwards on a larger scale in national housing programmes. The projects consisted of three pilot schemes concerning the design and construction of a new minimal cost community, the rehabilitation of old houses and the rational establishment and growth of spontaneous housing settlements.

To carry out the first pilot scheme in 1968 a competition was organized which was open to all Peruvian architects and to thirteen invited foreign teams. The competition was for the design and construction of a community of low-cost dwellings for 1,500 families on a 40 hectare site located 8 km north of Lima centre. The brief included very precise requirements as to layout and dwelling-types. Designers were invited 'to exercise inventiveness and experiment' to achieve a fairly high-density, compact housing, avoiding high multi-storied buildings.

All foreign entries were published in 1970 with extracts from the very comprehensive brief. The three international winners were Atelier 5; Herbert Ohl; Kikutaki, Maki and Kurokawa. It was, however, later decided to build a sample of the principal types entered by the different competitors including Christopher Alexander, Charles Correa, Candilis-Josic-Woods, James Stirling and Aldo Van Eyck, assisted by Sean Wellesly Miller.

a place which is their own . . .

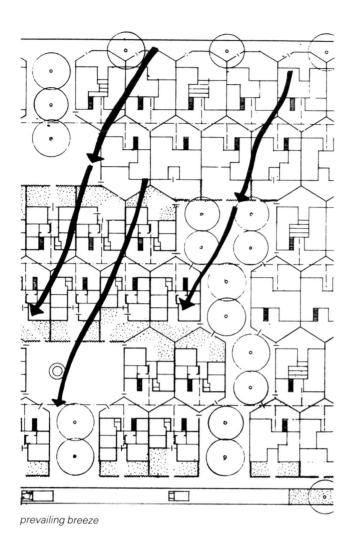

prevailing breeze

NO MISPLACED SUBURBIA

The general settlement layout is dealt with in some detail because the twenty-odd dwellings which were finally built do not demonstrate the full potential of the proposed clustering system. The urban structure for the entire site submitted for the competition is based on a clustering principle which is independent of either dwelling type or shape of lot. It thus possesses a general relevance beyond the given context and underlines the notion that rather than introducing endless variations of a **single** dwelling type, several altogether **different** types can go together within a single configurative urban fabric. The clustering bands are six lots deep, thus greatly reducing the number of parallel streets. Placed well apart, these streets — busy *avenidas* — serve both pedestrian and vehicular traffic, which **may well hurt** established planning ethics! However, in view of Peruvian urban reality, such a combined use of straight avenues (one kilometre long with a view of distant hills in both directions) need and should not be avoided. Whilst local traffic flow will be modest, its social and economic significance is considerable (e.g. *collectivo* taxis).

In contrast to the east—west, north—south traffic net, the pedestrian paths — *paseos* — run diagonally towards the wide middle avenue in the direction of the central plaza. All schools are accessible along these *paseos*. Access paths penetrate into the housing bands. They are wide enough to permit car entry for bringing building material to each dwelling when needed. Children's play places, trees for shade and fountains are situated at the ends of these access paths, whilst kindergartens are placed in the centre of the clustering bands. It is expected that small shops will develop spontaneously along the principal avenues.

. . .stage by stage with finally up to eight rooms after twenty to thirty years

NO MINIMAL DWELLINGS

Since barriada inhabitants do **not** end up with minimal houses (often up to eight rooms after twenty—thirty years), an attempt was made **to avoid the ethos of the industrially produced minimal house**, which, besides bearing the stigma of poverty, can so easily penalize a family in later life. The design guarantees ample scope, allowing it to respond to the social dynamics of the Peruvian working class.

Like the typical barriada house, this one can expand, through self-help or otherwise, horizontally or vertically from one to eight rooms in the course of time and according to a family's requirements, endeavour and resources.

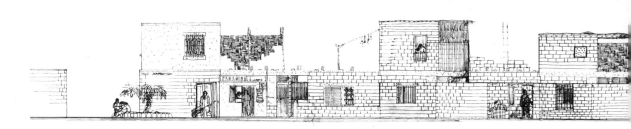

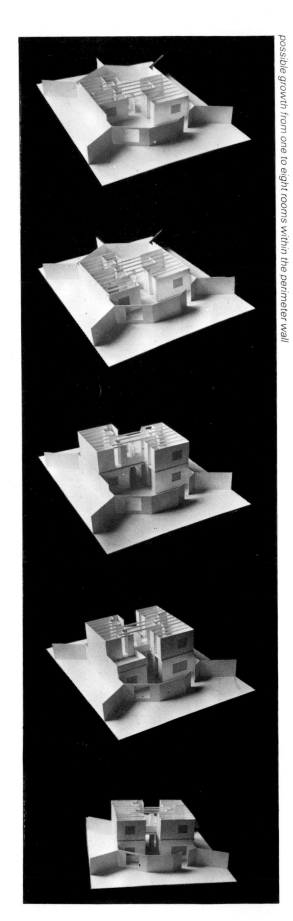

possible growth from one to eight rooms within the perimeter wall

House grouping, ground...   and first floor plans

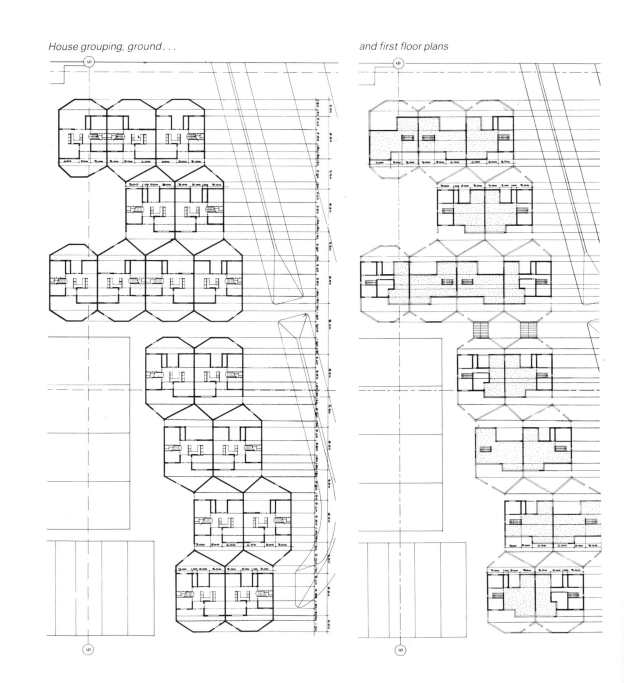

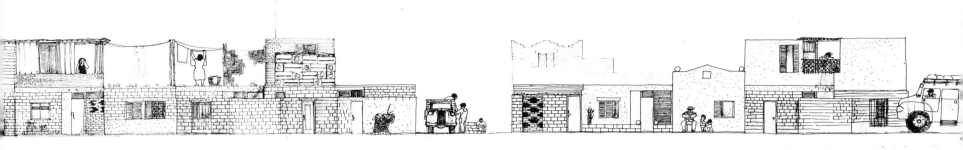

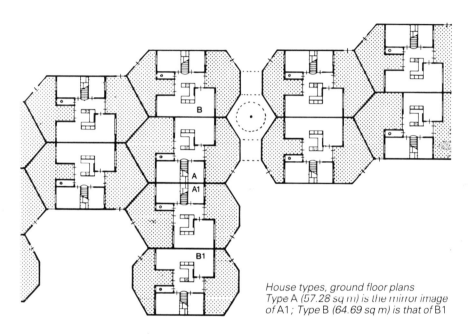

House types, ground floor plans
Type A (57.28 sq m) is the mirror image
of A1; Type B (64.69 sq m) is that of B1

NO MINIATURE PATIOS

Since small patios so often provide **four walls for the next roof**, resulting in the loss of outdoor space, direct access to all rooms and grave reduction of light and air in surrounding rooms – even complete loss of both when lots adjoin along two or three sides which is usually the case. Instead each house has an enclosed front and back yard (or garden) connected by a central space which can be partially or entirely opened on to either one or both for ventilation and free movement. The sawtooth non-loadbearing yard walls discourage expansion outside the house's maximum orthogonal perimeter, i.e. into the yards. **That future free development should not work against the best interests of the occupant is still insufficiently realized when designing expandable dwellings!** The central multi-interpretable outdoor-indoor space which runs somewhat diagonally through the middle of the lot from end to end contains the kitchen and gives access to all rooms and roof terraces. Prevailing wind direction (SSW) called for approximately east to west placing of rows; whilst high all-year-round humidity required an open, fairly loose structure. The deep access paths to the dwellings allow breezes to penetrate far into the clustering bands. The lots are shifted laterally, facilitating the penetration of air through the houses.

Aerial view of the project under construction

typical greenery and additions

241

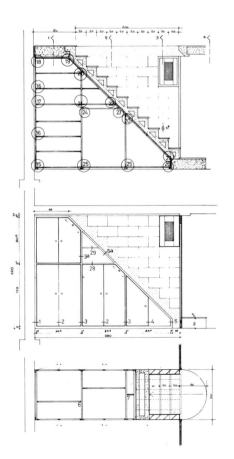

Left *the staircase spans in parallel parts from top to bottom for the sake of free storage space underneath*

windows of local construction using small panes of glass to lower replacement costs

Opposite page *elevations, sections and plans of a selection of house types*

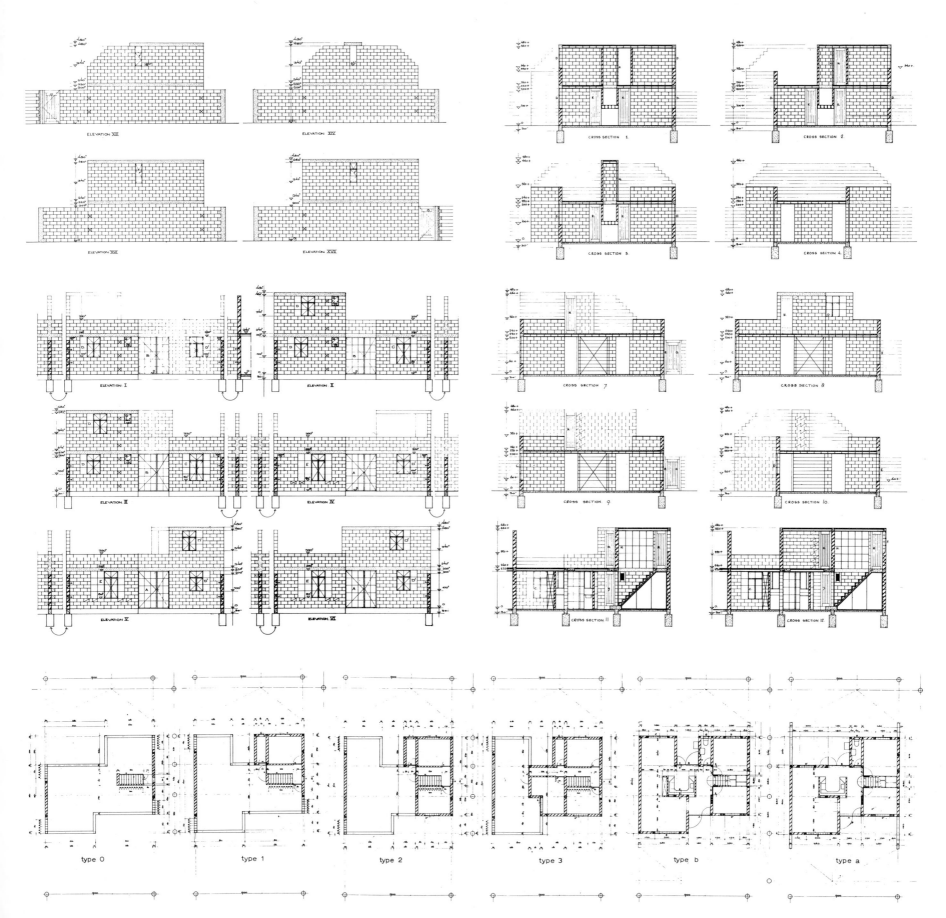

ELEVATION XII

ELEVATION XIV

ELEVATION XVI

ELEVATION XVII

CROSS SECTION 1.

CROSS SECTION 2.

CROSS SECTION 3.

CROSS SECTION 4.

ELEVATION I

ELEVATION II

ELEVATION III

ELEVATION IV

ELEVATION V

ELEVATION VI

CROSS SECTION 7

CROSS SECTION 8

CROSS SECTION 9

CROSS SECTION 10.

CROSS SECTION 11

CROSS SECTION 12

type 0

type 1

type 2

type 3

type b

type a

central space with kitchen...

connecting front and back courtyards...
when the doors are open a single continuous external living space is formed

in humid Lima a breezy roof is a better outdoor space than a patio

NO FALSE CHOICES

It was felt that the rigidity of form and norm, which unfortunately still characterizes conventional low-cost housing, is particularly unsuitable to the needs of a poor immigrant urban population with unpredictable social and economic mobility adapting to a new life-style and living pattern.

This house type assumes that what people have come to want today (e.g. a kitchen not visually connected with the living room) is not necessarily what they may want in the future and hence should not form an unchanging basis of design. However, it would be equally wrong to impose *new* or *advanced* solutions wholesale, however reasonable they may seem, **if they go against the grain** of existing sentiment. Hence the dwellings, whilst responding to current norms, at the same time provide scope for changing aspirations and sentiments, the nature of which cannot be safely predicted but will probably move towards a more open form of living within the house. The question of existing norms and forms and the ways these will change confronts the architect with an apparent choice **which it should not be his concern to make**. The line beyond which improvement becomes interference, imposition and impediment is not easy to draw – nor is it one which architects like drawing!

Although not mandatory, the competition programme clearly stimulated the development of new prefabrication or industrialized building methods, for very obvious reasons. However, we declined the use of materials or constructions not currently available or used by people with limited building experience and equipment.

Painting the houses should have been left to the people as specified, but the *client* thought differently. **Now the whole place is like a post-Second World War Third World Weissenhof Siedlung – white all over![7]**

The entire settlement remained unoccupied years after completion. Finally all the dwellings were sold to middle-income families, thus bearing out a suspicion that was with us from the beginning.

7 Since that was written the colours are coming freely.

MOTHERS' HOUSE, AMSTERDAM

The Mothers' House was built in an urban setting for a client whose programme was both explicit and wise. This made the task of finding the **right** form to complement it particularly worthwhile.

TRANSPARENCY AND ENCLOSURE

I have been busy for some years now revaluing the notion of transparency in the light of that other notion – enclosure – convinced as I am that architecture misses the mark and evades its purpose by reverting in turn to the one at the cost of the other for little more than stylistic reasons. So the fact that my client – an admirable one if ever there was one – desired an **open house** came just at the right moment, nourishing a notion already growing in my mind: that it would be expedient, both in this particular case and in general, to bypass trying in vain to arrive at the right kind of openness (which presupposes the right kind of enclosure and vice versa) **in spite of** transparency.

What is due, now, is to move step-by-step towards **enclosure brought about** not in spite of, but by means of **transparency**. Not for stylistic reasons, no, no, no, for **style** comes as a reward, but for what transparency can still provide on a human level. I say human level intentionally because, with or without RPP (Rats, Posts and other Pests)[8] and their counter-humanist stance, **there is no other level**. (How ill-founded their renunciation of the modern movement is, the astounding number of gratifying buildings built within a few decades shows.)

To return from temporary darkness (the RPP blight) to what Robert Delaunay once called 'l'ombre des ultra-violets' and from there on to the rainbow in the light of which this building's transparency was articulated in **depth**, the necessary **enclosure** defined and the desired **openness** brought about.

OPEN WHAT WOULD OTHERWISE CLOSE

What people so diligently subtract from virgin – open – exterior space with the help of material and construction all too often 'closes' in the process. Instead of being 'interiorized', it is rendered empty and emptiness excludes accessibility. That is why the recent trend towards emptiness (the perforated solid – Rossi) is so very vicious. It is precisely because constructing in exterior space inevitably entails demarcation, enclosure, separation and size reduction that I now wish to place special emphasis on a quality which does **not** survive all delimiting activity without conscious effort: openness. Openness, therefore, requires our utmost attention. It is in fact the very thing architects in particular are required to maintain: to reconstitute by means of construction, **thus keeping open what (without their special care and competence) would otherwise close**.

So open – keep opening – what, if you fail to do so, will sooner or later cease to be space altogether. In the face of such a ghastly prospect, never forget that space can still bring light and light reveal enclosure.

8 Rationalists, Post-Modernists and Other Pests.

THE COLOURS – THE RAINBOW

Steel and paint are closely allied: one tends to forget this, taking it for granted. Ships, railway engines, motor cars, bicycles, bridges – a host of things – are painted and repainted for protection according to custom, tradition or, if they happen to be pipes like those which run, up, down, along and across the Beaubourg, just for fun: so where there are no pipes there is no fun!

In my case the moment arrived, when once again a building for the most part of exposed steel, was to be painted without there being any known or accepted way – without experience, knowledge or more than sporadic examples to show what active colour, an unquestionable ingredient of environment in general, means in terms of architectural space in particular. Miesian steel tends towards blacks but my steel is anything but Miesian! Now when I think of active colour in buildings or cities, it is not gaudy signs, advertisements, vehicles and throw-away materials which I see before me, for these alone, bar some clothes and an occasional front door, represent active colour. No, I see rainbows – ones which remain after it has stopped raining or the sun has gone. But when I think of paint and active colour together, I think of painters first. And when I think of them I naturally see paintings in my mind's eye – the results of thought, experience, discovery and artistry. So, via a kind of logic a little different no doubt and with a steel building standing before me as yet unpainted, I tried to think of it as a painting, ending up painting it as though it actually was two-dimensionally spatial!

Not in front of me but wrapped around me. As if I were **immersed** in it, to borrow Cézanne's wonderful term. In front the twelve colours follow the hollowing out in spectral sequence from blue back to blue. The red of the interior partition walls appears where one of them on the top floor becomes 'exterior'. I spent almost as much time and thought on the colours as I spent on the building without them! Trying again and again I do think I managed to get a little way beyond the beginning.

What architecture needs if it is to become really **useful**, to make sense, is at least a few people with the kind of supreme intelligence, vision, patience and consummate artistry Seurat, Cézanne and Mondrian had. In the world of painting, there were so many – so many and for so long! It is of course not because of the colours alone that I mention this.

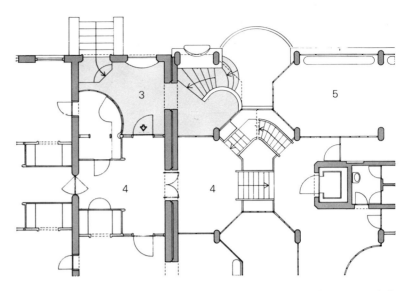

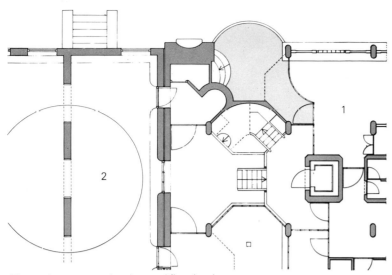

Plans at lower ground and ground floor levels.
1 Bicycles and prams 2 Playroom 3 Entrance 4 Hall 5 Cafeteria

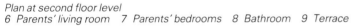

Plan at second floor level
6 Parents' living room 7 Parents' bedrooms 8 Bathroom 9 Terrace

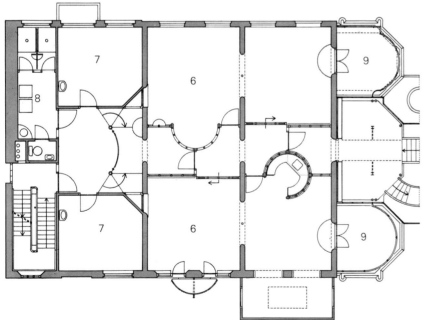

It is those **other** requirements – the mute kind – which seldom speak from a client's brief that can best goad an architect, sensitive to them, to the design-idea which appropriately fulfils the verbalized requirements which do speak from the brief.

Buildings, especially within the fabric of old city centres, that misbehave towards what is already there; towards what exists outside, will also misbehave towards what is inside, towards what they are expected to serve in the first place. In short, they must also be **right** irrespective of their particular – specified – function. The relationship between a building's interior spatial organization and the exterior urban setting to which it is added and one hopes will manage to belong, is not only a formal and spatial one, it is also a temporal one. That is why buildings should have temporal perspective and associative depth – so very different, both, from arbitrary historicizing indulgence!

A building doesn't need to **look** like its neighbours in order to avoid alienation through incongruity, but it should take heed nonetheless. **Response through mimicry is faint-hearted and in every way futile**. That is why the 'typology' addicts of today are such a liability!

Here, in this building, extension and conversion form a combined process. Entry takes place where old and new meet. The vestibule of the old house becomes an external porch – accessible from the new via some steps, it in turn gives access to the new via the old. Conflict between the existing home and its new extension is resolved at the entrance – an in-between realm (also in a temporal associative sense). The irregular floor levels of the old building are extended into the part of the new one immediately adjacent to it, so that the 'split' with the new regular levels is shifted away from the party walls between the two; like the portico, another unifying device, since this is where the stairs connecting the floor levels of both buildings occur. By reducing the floor area upwards from floor to floor, i.e. stepping back the volume, sun and air penetrate deep into the building, while loggias and roof terraces result. Buildings in general (those along an urban street no less) should be populated externally, so that nobody feels tucked away behind walls and windows, cut off from the world outside.

Especially in a home like this one, something more generous than balconies is required. A glass-roofed internal 'street' feeds the five children's departments from above, bringing extra light into the rooms. After dark, parents watching from above see movement along it . . . and vice versa.

249

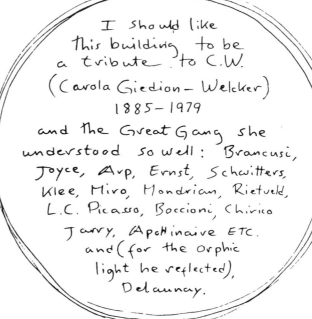

I should like
this building to be
a tribute to C.W.
(Carola Giedion—Welcker)
1885—1979
and the Great Gang she
understood so well: Brancusi,
Joyce, Arp, Ernst, Schwitters,
Klee, Miro, Mondrian, Rietveld,
L.C. Picasso, Boccioni, Chirico
Jarry, Apollinaire ETC.
and (for the Orphic
light he reflected),
Delaunay.

A.v.Eyck

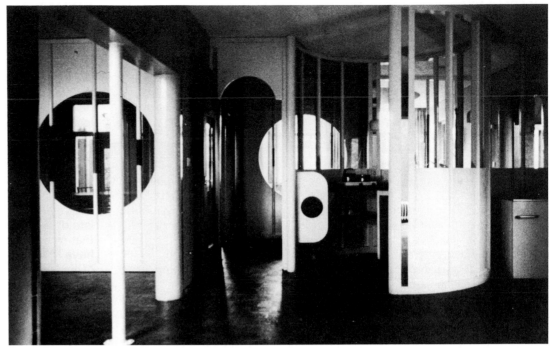

ONLY ALICE

Tiles outside – tiles inside. The partly spectral tile-tableaux (red – orange – green – yellow – blue) with mirrors all round, set in four concrete panels along the street, are repeated in the toilets without change of height (5 × 15 = 75cm); only a reduction in width forming single narrow (4cm) vertical strips. Thus, between the **largest space outside and the smallest inside** – between street and WC – the **intensity** changes but not the **scale** – a device out of many to keep the latter in hand – constant – through careful local dimensional adjustment. Small elaborations of this kind ('elaborations' rather than 'decorations') were introduced in more places for similar reasons. **Only Alice, after all, grows larger and smaller in turn**. The rest of us are left to witness how, albeit through our own doing, what is there around us expands and contracts at random with nervewracking abruptness; always either too large or too small – in short, always more or less the wrong size – outsize! Characteristic of any good door, window, building or urban space is that they possess what I call **right-size** (many sizes at once though **one** scale).

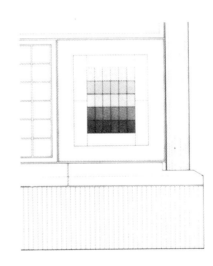

tiles outside . . .

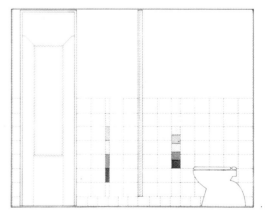

. . . tiles inside

PHOTOGRAPHIC CREDITS AND ACKNOWLEDGMENTS

All the illustrations in this book have been provided by the individual contributors. Where possible the photographic sources have been identified as listed and thanks are due to these photographers and to those whose identity it has not been possible to trace.

Architectural Review	38, 39
David Barrable	139, 141
Behr Photography	137
John Bethel	34, 35, 37
Jerry Bragstad	33
Brecht-Einzig	86 (Kiruna), 88, 136 (upper), 142 (lower), 146, 147, 150, 193, 194, 195 (Olivetti)
Council of Industrial Design	178
Violette Cornelius	236, 237, 239
Gordon Cullen	186 (drawing)

Rolf Dahlström	80, 93
Richard Davies	114, 122, 123, 128
Department of the Environment	203
Director General of the Ordinance Survey	44 (map)
Tony Dobbin	151
John Donat	114, 117, 134, 144, 145, 151, 152, 153, 156, 158, 162, 163, 164, 195 (Art Centre & Museum), 200, 202
Y. Futagawa	230, 231, 232, 233

Peter Hall	137 (Epidauros), 208, 210
P. Hester	208, 210
Peter Land	242
Edward Leigh	160
The Daily Mail	81
John Murphy	135
Donald Mill	138, 140, 141, 142 (upper), 143, 148, 149
Lars Mongs	91 (exterior)
David Moore	220, 221
Alberto Ponis	136 (lower)

Alberto Rogas	235
David Shalev	94, 98, 99, 102-106
Tony Smith	92 (Clare Hall)
Sundahl	78, 82, 86 (Shopping Centre), 91 (interior), 92 (Church), 214
Bill Toomey	39
W. J. Toomey	182
Peter Walser	198, 199
Arno Whalberg	90
Strichting Wonen/G.v.d. Vlugt	253

INDEX